CAREERS
IN ENGINEERING

VGM Professional Careers Series

CAREERS
IN ENGINEERING

Geraldine O. Garner

VGM Career Horizons
a division of *NTC Publishing Group*
Lincolnwood, Illinois USA

Cover photo courtesy of Universal Oil Products

Library of Congress Cataloging-in-Publication Data

Careers in engineering / edited by Geraldine O. Garner.

p.   cm. — (VGM professional careers series)
Includes bibliographical references.
ISBN 0-8442-4184-9 — ISBN 0-8442-4185-7 (pbk.)
1. Engineering—Vocational guidance.   I. Title.   II. Series.
TA157.G32   1993
620′.0023′73—dc20

92-45806
CIP

1995 Printing

Published by VGM Career Horizons, a division of NTC Publishing Group.
© 1993 by NTC Publishing Group, 4255 West Touhy Avenue,
Lincolnwood (Chicago), Illinois 60646-1975 U.S.A.
Manufactured in the United States of America.

5 6 7 8 9 0 VP 9 8 7 6 5 4 3 2

CONTENTS

What licensing can mean for your career.   Licensing
requirements.   Tips for meeting the licensing
requirements.   Additional information.

Statistics on women in engineering.   Advice to women
entering engineering.   Opportunities for women in
engineering.   Scholarship information.   Organizations
for women in engineering.   Additional reading.

Statistics on minorities in engineering.   Advice for
minority engineers.   Salary statistics for minority
engineers.   Organizations for minority engineers.
Other sources of information.   Additional reading.

ABOUT THE AUTHOR

Geraldine O. Garner is currently assistant dean and director of the Walter P. Murphy Cooperative Engineering Education Program at Northwestern University's McCormick School of Engineering and Applied Sciences. Prior to coming to Northwestern, she taught graduate and undergraduate courses in career development at Virginia Commonwealth University.

Dr. Garner received both her B.A. and M.Ed. from The College of William and Mary and her Ed.D. in Career Counseling from Virginia Tech. She is the author of a variety of books, articles, and papers, and has received numerous honors throughout her career.

FOREWORD

Engineering is a unique profession. The choices incoming students have between engineering fields are both numerous and diverse. Building bridges requires structural engineering knowledge; automobile design needs mechanical engineers; and communications technicians must be trained in electrical engineering. As we attempt to recapture the country's former status as the world's technology leader, today the United States is in need of talented young engineers in every corner of industry.

The modern engineer has the benefit of viewing the history of past engineering feats that serve both as testimony to great minds and artifacts of incredible achievement given the limitations of centuries past. Today the computer and its artificial intelligence abilities aid the engineer in his or her tasks, increasing efficiency in design, safety, and overall effectiveness.

Those who aspire toward engineering degrees and work in a diverse field have a tough school workload ahead of them. But the pinnacle of personal accomplishment and satisfaction await the successful. Here is a world where design, mechanics, and craftsmanship can be invented, brought together, and honed to create something as small as a computer chip or as large as a nuclear-power aircraft carrier, from a simple time-lock mechanism to the complexities of the Space Shuttle. Here is your chalkboard—where imagination meets reality.

The Editors
VGM Career Horizons

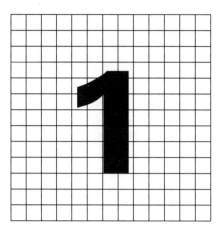

WHAT IS ENGINEERING?

Perhaps the best way to answer the question "What is engineering?" is to look at the definition of engineering given by the American Society for Engineering Education.

> Engineering is the profession in which a knowledge of the mathematical and natural sciences gained by study, experience, and practice is applied with judgment to develop ways to utilize the materials and forces of nature, economically for the benefit of mankind.

A HISTORY OF ENGINEERING

In order to gain a better perspective on the engineering profession, it is useful to understand how the profession developed through the centuries.

The first engineer known by name and achievement is Imhotep, who built the famous stepped pyramid in Egypt circa 2550 B.C. The Persians, Greeks, and Romans, along with the Egyptians, took engineering to remarkable heights by using arithmetic, geometry, and physical science. Many famous ancient structures that are still standing today demonstrate the ingenuity and skill of these early pioneers of engineering.

Medieval European engineers, like the ancient engineers, combined military and civil skills to carry construction, in the form of the Gothic arch and flying buttress, to heights unknown by the ancients. The sketchbook of Villard de Honnecourt, who lived and worked in the early thirteenth century, demonstrates the Gothic engineer's remarkable knowledge of natural and physical science, mathematics, geometry, and draftsmanship.

In Japan, China, India, and other Far Eastern areas, engineering developed separately but similarly. Sophisticated techniques of construction, hydraulics, and metallurgy practiced in the Far East led to the impressive, beautiful cities of the Mongol Empire.

In 1747, the first use of the term *civil engineer* coincided with the founding in France of the first engineering school, the National School of Bridges and Highways. Its graduates researched and formalized theories on many subjects, including fluid pressure. John Smeaton (1724–1792), British designer of the Eddystone Lighthouse, was the first to actually call himself a civil engineer, thus separating his work from that of the military engineer. The eighteenth century also witnessed the founding in Britain of the world's first engineering society, the Institution of Civil Engineers.

Civil engineers of the nineteenth century designed sanitation and water-supply systems, laid out highways and railroads, and planned cities. Mechanical engineering had its beginnings in England and Scotland and came out of the inventions of Scottish engineer James Watt and the textile machinists of the Industrial Revolution. The rise of the British machine-tool industry caused interest in the study of mechanical engineering to skyrocket, both in Europe and abroad.

The gradual growth of knowledge in the area of electricity in the nineteenth century eventually led to the most popular branch of the engineering profession—electrical and electronic engineering. Electronics engineering came into prominence through the work of various English and German scientists in the nineteenth century and with the development in the United States of the vacuum tube and the transistor in the twentieth century. Electrical and electronics engineers now outnumber all other engineers in the world.

Chemical engineering came into existence through the nineteenth-century spread of industrial processes involving chemical reactions to produce textiles, food, metals, and a variety of other materials. By 1880, the use of chemicals in manufacturing had created a new industry, mass production of chemicals. The design and operation of this industry's plants become the main function of the new chemical engineer.

The twentieth century has brought many other branches of the profession into prominence, and the number of people working in the engineering field has increased dramatically. Artificial hearts, airplanes, computers, lasers, plastics, space travel, nuclear energy, and television are only a few of the scientific and technological breakthroughs that engineers have helped to bring about in this century. No doubt the field will continue to grow and expand into the next century.

THE TECHNOLOGY TEAM

It is evident that engineers address the challenges that face the society in which they live. From the Egyptian pyramids and the compounds for medieval swords to high-definition television and the aerospace plane, engineers are above all problem solvers. They link scientific discovery with day-to-day applications.

Engineers are team players who improve products, processes, and services. Therefore, it is important to understand the technology team on which engineers participate.

The engineer is a part of a team of specialists whose goal is to apply scientific knowledge and practical experience to the solution of technical problems. This "technology team" is a work force consisting of scientists, engineers, technicians, and craftsworkers. Everyone on the team works together to solve a problem or to invent a useful device or system. In learning about the makeup and function of the technology team, you can develop an understanding of technology as a whole and of how it is put to use.

At one end of the spectrum of technology is the *scientist*. The scientist's purpose is to discover knowledge. He or she seeks to uncover new facts and to learn more truths about the natural world. Furthermore, scientists seek to explain the facts that they discover by developing new theorems or theories which relate causes and effects in the natural systems they have investigated. In their work, scientists seek to *know* rather than to *apply*. In other words, their principal concern is not the application of the new knowledge they have discovered but simply the discovery of that knowledge itself. Some scientists are interested in developing applications of science and scientific methods, but the principal activity of even these individuals remains the discovery of new knowledge.

The *engineer*, in contrast to the scientist, is interested primarily in the application of scientific knowledge about the natural world and in discovering facts about the artificial world created by humans. The primary responsibility of engineers, as a part of the technology team, is in conceiving and planning efforts to apply scientific knowledge. They design and plan developmental projects, production processes, operations and maintenance procedures, and so on. Their activities are devoted to designs and plans to achieve certain results. These results almost always benefit society; however, the purpose beyond that is to achieve this benefit at minimum cost in money, materials, and time. In an effort to achieve efficient results, the engineers attempt to forecast the behavior of a system they have designed or to predict the accomplishments of a planned program. All benefits and costs of proposed activities must be predicted by engineers, who are the principal planners of the technology team.

It is the *technician's* responsibility to see that the engineer's design or plan is implemented. While the engineer is concerned mainly with designing or conceiving, the technician is concerned with doing. The technician may be involved in time-and-motion studies or in supervising the construction of a facility designed and planned by the engineer. In accomplishing such work, the engineering technician is more specialized and more concerned with a particular application of scientific knowledge than is the engineer who must plan complex systems. Basically, the technician utilizes science and mathematics to solve technical problems contained within the broad framework or designs and plans conceived by an engineer. Additionally, he or she utilizes instruments and certain tools to measure and monitor the quality and performance of completed systems. However, the technician's principal function is not to utilize tools but to see that designs and plans are implemented by the craftsworkers who do use tools. The technician lies in the occupational spectrum closest to the engineer.

At the opposite end of the technology spectrum from the scientist is the skilled *craftsworker*. Craftsworkers use their hands and special skills rather than science or scientific knowledge. They are more likely to employ tools than instruments in their work, and they must develop a high degree of skill in using these tools. Craftsworkers include electricians, instrument makers, machinists, model makers, and others. The craftsworker, too, has an important position on the technology team, and, to some degree, the overall success of the technical system depends upon his or her skill in utilizing tools and his or her concern for good workmanship in construction.

WHAT AN ENGINEER DOES

Engineers plan, design, construct, and manage the use of natural and human resources. Some would say that engineering is both art and science. In addition to human skill, engineering also involves science, mathematics, and aesthetics. Engineers solve problems. They design cars, spacecraft, and medical devices; they build buildings and bridges; they solve environmental problems; they apply computer technology to a wide range of problems.

Because engineers have a strong interest and ability in science, mathematics, and technology, they are team leaders who can take an idea from concept to reality. The professional societies in engineering recognize more than 25 specialties and over 85 subdivisions within these specialties.

There are seven major functions common to all branches of engineering.

Research.   A research engineer looks for new principles and processes by using scientific and mathematical concepts, by experimenting, or by using inductive reasoning.

Development.   A development engineer takes the results of the research and puts it to use. Creative and intelligent application of new ideas may give the world a working model of a new machine, chemical process, or computer chip.

Design. A design engineer chooses the methods and materials necessary to meet technical requirements and performance specifications when a new product is being designed.

Construction. A construction engineer prepares the construction site, arranges the materials, and organizes personnel and equipment.

Production. A production engineer takes care of plant layout and the choosing of equipment with regard to the human and economic factors. He or she selects processes and tools, checks the flow of materials and components, and does testing and inspection.

Operation. An operating engineer controls manufacturing and processing plants and machines. He or she determines procedures and supervises the work force.

Management. Engineers in the management area analyze customer needs, solve economic problems, and deal in a variety of other areas depending on the type of organization involved.

Even within the different branches of engineering, there is no one generic engineer. As we see above, there exists a wide variety of areas in which the prospective engineer can find satisfying and rewarding work.

In addition to diversity of function, engineering is also performed in a wide variety of private, commercial, and government settings. Many engineers are found in manufacturing industries but they are also found in nonmanufacturing settings such as banks and hospitals. In addition, they work in engineering and architectural firms; public utilities; business and management consulting firms; federal, state, and local governments; and colleges and universities.

THE EDUCATION AND TRAINING OF ENGINEERS

The National Society of Professional Engineers recommends that future engineers take:

- Algebra I & II
- Geometry
- Trigonometry
- Calculus
- Biology
- Chemistry
- Physics
- English (4 units)
- Social Studies (3 units)
- Foreign Languages (2–3 units)
- Fine Arts Humanities (1–2 units)
- Computer Programming or Computer Applications

Other courses that may be helpful include economics, history, and public speaking. It is also recommended that prospective engineering students take advanced placement (AP) or honors level courses and set a goal of achieving combined scores of at least 1000 on the SAT exam or 20 on the ACT exam.

Admissions officers at engineering colleges and universities also look for well-rounded students. Extracurricular activities during high school can reflect this. Being a member and holding an office in math and science clubs will demonstrate strong and consistent interests related to engineering. However, participation in athletics, service organizations, and cultural activities are also important.

Bachelor's degree programs in engineering are available through colleges or universities accredited by the Accreditation Board of Engineering and Technology (ABET). These programs take four or five years. In general, the first two years concentrate on mathematics and the physical sciences with introductory engineering courses and courses in English and the social sciences. The last two years include required courses in engineering and particularly required courses in the major. In addition, engineering students take technical electives and "free" electives. Free electives are either any course outside of engineering that the student wishes to take or any course on a list of approved electives by the engineering department.

There are also transfer programs called "Two-plus-Two" or "Two-plus-Three" programs. These programs combine two years of study at a community college and then two or three more years of study after transferring to a participating four-year college. In some cases, a bachelor's and master's degree is awarded at the end of a "Two-plus-Three" program.

There are also five- or six-year cooperative engineering education programs. In these programs, engineering students alternate periods of academic study with periods of paid engineering-related work in industry.

In an accredited engineering program students are required to spend the first two years studying basic sciences: mathematics, physics, chemistry, and introductory engineering. They also take humanities, social sciences, and English. During the last two years, most courses are in engineering. It is during this time that students can elect to specialize in one area within their discipline or complete a general engineering degree in their chosen area of engineering.

CAREER PATH SCENARIOS

A bachelor's degree in engineering provides a wide range of career options. In addition to employment opportunities in industry, business, and government, engineers are employed in such areas as production, sales and marketing, management, and research and development.

Many engineering graduates decide to pursue graduate and professional degrees. While some obtain master's and doctoral degrees in engineering, others decide to pursue graduate and professional degrees in business administration, medicine, and law.

Advanced degrees offer new opportunities for career advancement in business and industry. They also open doors for college teaching and research careers.

FUTURE PROJECTIONS

Many factors in today's global economy suggest that the future projections for engineering are excellent. This is especially true in engineering areas such as chemical engineering, environmental engineering, materials engineering, and manufacturing engineering.

However, engineering opportunities have always had a tendency to go in cycles. Aerospace engineering is a good example, as is chemical engineering. The changing economy and world events impact opportunities for engineering. During the remainder of the 1990s and beyond the year 2000, there will be a shift from defense industries, which have provided opportunities for many engineering disciplines during the last 50 years, to new industries and emerging industries that will allow us to compete in a global market. The health care industry is an example of a growing industry that is providing new opportunities for engineers.

Federal and state governments regularly publish documents that provide current and up-to-date information on the future outlook and projections for a wide range of career fields. *The Occupational Outlook Handbook* published by the U.S. Department of Labor is a good source of information in this area. *The Guide for Occupational Exploration*, which is also produced by the U.S. Department of Labor, provides a list of questions that are designed to help assess interest in engineering. Both publications are available from the U.S. Government Printing Office and in most guidance offices and public libraries.

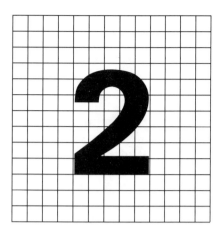

MAJOR ENGINEERING AREAS

Although the U.S. Department of Labor lists 25 engineering specialties with at least 85 different subdivisions, it is important to keep in mind that there are six areas of engineering that form the core of the profession. Preparation in any one of these areas will provide a solid foundation for a wide range of specialty areas. These six areas are chemical engineering, civil engineering, electrical engineering, industrial engineering, materials engineering, and mechanical engineering.

Undergraduate study in any one of these areas will be adequate preparation for many career options after graduation, including graduate study in the same area or another area of engineering. Study of one of these basic areas of engineering is also excellent preparation for the study of business, law, or medicine. In a time when it is important to keep all options open for future career development, pursuit of one of these disciplines can provide flexibility and satisfaction throughout one's professional life.

CHEMICAL ENGINEERING

Chemical engineers combine the science of chemistry with the discipline of engineering in order to solve problems and find more efficient ways of doing things.

Chemical engineers are involved in many phases leading to the production of chemicals and chemical by-products. Engineers design equipment and chemical plants as well as determine the best methods of production. Thus, chemical engineers can be considered combinations of industrial engineers, materials engineers, electrical engineers, and chemists.

The Nature of the Work   Although one of the smaller engineering disciplines, chemical engineering is one of the oldest and most established. In combining the science of chemistry with the discipline of engineering, chemical engineers deal with such problems as:

- developing more efficient methods of refining petroleum
- purifying polluted water and air
- developing more durable and versatile products such as plastics
- harnessing solar and geothermal sources of energy
- recycling reusable metals, glass, and plastics
- producing cheaper and better fertilizers and pesticides
- creating more effective paints, dyes, and coatings
- manufacturing improved electronics/semiconductors
- producing safer cosmetics and more effective pharmaceutical products

Chemical engineers are largely responsible for the foods we eat, the fuel that supplies our needs, the control of pollution, and the recovery and use of raw materials from our oceans.

While chemists are interested in the basic composition of elements and compounds found in nature and seek to invent new products from them, chemical engineers work to develop these new products and to evaluate them practically and economically. Chemical engineers work in many of the following areas.

Research and Development.   Chemical engineers in this area spend much of their time designing and performing experiments and interpreting the data obtained. They invent and create new and better ways of developing products, controlling pollution, reducing safety and health hazards, and conserving natural resources. Their findings may be refined in laboratories but more often in a pilot plant, which is a miniature version of the proposed commercial facility.

Some research and development (R&D) is conducted by industry to develop improved products. Most research, however, is conducted in universities under the direction of college faculty with the assistance of graduate students. Research is aimed at technology development and a better understanding of the different ways chemicals can interact under various conditions.

Design and Construction.   Project engineers design and construct chemical manufacturing facilities. They may work directly for a manufacturing firm or for a consulting company hired by the manufacturer.

In design work, chemical engineers draw heavily on their knowledge of mathematics, physics, chemistry, and other related sciences. They use this knowledge to select and size equipment and determine the optimum method of production. Control systems are designed to maintain consistent product quality, minimize waste generation, and assure safe operation of the facility. They develop capital and operating costs and present anticipated profitability statements to justify the proposed project. After the project is accepted, detailed specifications, drawings, and priority schedules are prepared.

Chemical engineers in design and construction may act as field engineers directing and assisting workers during the construction period. After construction, these engineers may assist in equipment testing, operator training, and plant start up.

Operations/Production.   Chemical engineers in operations are responsible for the day-to-day operation of a manufacturing facility. Their primary interest is in the production of a product economically and safely in order to meet the customers' needs in both quality and quantity.

They are challenged by raw material variations and shortages, labor disruption, cost fluctuations, weather, and equipment breakdowns. They gradually adjust operating conditions to achieve improved product yield and quality and reduced operating costs.

Technical Sales.   Chemical engineers involved in technical sales not only have strong technical skills, they also have exceptional people skills. It is their responsibility to introduce new products to customers and to assess why some products do better than others in the marketplace. In the area of total quality management, chemical engineers involved in technical sales provide a vital link in determining why a product is not functioning to a customer's satisfaction.

Environmental and Waste Management.   Chemical engineers employed in this area devise techniques to recover usable materials from waste products and develop methods to reduce the pollution created during the manufacturing of a product. They also design waste storage and treatment facilities as well as design pollution control strategies for plant operations.

The Settings in Which Chemical Engineers Work

The chemical industry is the primary employer of chemical engineers. Because chemical engineers are primarily concerned with the large scale manufacture of products from raw materials through closely controlled physical and chemical changes, they find employment in such diverse areas of the chemical industry as agricultural chemicals, plastics, and industrial chemicals.

The petroleum industry as well as the pharmaceutical, cosmetic, and food processing industries also employ a significant number of chemical engineers. In fact, the food processing industry has received increasing attention in recent years with the growing need for food to nourish the ever-expanding world population.

Other emerging industries for chemical engineers include atomic energy development and coal conversion. However, the fastest growing area for chemical engineers is the environmental industry. Government agencies, such as the Department of Energy and the Environmental Protection Agency, also employ a large number of chemical engineers.

Chemical engineers work in local, state, and federal government agencies to advise lawmakers on environmental issues and industrial concerns. They develop laws and standards to protect the environment and the public from chemical hazards.

Other Settings.   In addition to these areas, there are others. For example, almost one-third of the chemical engineers in the United States function as managers and supervisors. Still others are teachers at universities where chemical engineering education is provided.

Training and Other Qualifications

Education for chemical engineering begins in junior high school with the appropriate math and science courses as preparation for three years of high school science, including chemistry and physics, four years of mathematics through trigonometry or calculus, and at least three years of English. Education continues in college with a four-year degree in chemical engineering. Positions in teaching or research also frequently require additional college training at the master's and Ph.D. level.

The American Institute of Chemical Engineers cooperates with the Accreditation Board for Engineering and Technology to accredit undergraduate curricula in chemical engineering. A current list of those programs can be obtained by writing to the American Institute of Chemical Engineers. According to the American Institute of Chemical Engineers, students pursuing degrees in chemical engineering can expect to take the following courses in college:

Basic science: Math, physics, chemistry, and some life sciences.

Engineering science: Material and energy balances, thermodynamics (dealing with heat), control of chemical processes and fluid mechanics (flow of liquids and gases), mechanics of solids, electrical and electronics engineering, materials science, and computer science.

Engineering design: Engineering economics, design of chemical reactors, heating and cooling apparatus, and piping.

Communications: English, speech, technical writing, computer languages, and graphics.

General education: History, philosophy, psychology, sociology, anthropology, art, music, and literature.

Outlook for the Future

The continued demand for new and improved products and more economical processes will make for an exciting, dynamic future for chemical engineers. In addition, chemical engineers who initially join federal or state environmental protection agencies to gain several years of experience can easily move into the private sector. Their engineering knowledge as well as their understanding of environmental regulations will be highly sought after.

Earnings

According to the 1992 Lindquist-Endicott Report, the average monthly starting salary for a person graduating with a BS degree in chemical engineering was $3,294.00. For those graduating with a master's degree in chemical engineering the average monthly starting salary was $3,506.00. New Ph.D.'s in chemical engineering reported an average monthly starting salary of $4,667.00.

Additional Sources of Information

American Institute of Chemical Engineers
345 East 47th Street
New York, NY 10017

The American Chemical Society
1155 Little Falls Street
Falls Church, VA 22041

CIVIL ENGINEERING

Of all the engineering professions, civil engineering is recognized as the oldest and the broadest of the engineering disciplines. Civil engineering extends across many technical specialties that interact with one another. As a result, civil engineering is the largest single user of high technology including computer-aided design for projects that range from rural roads to launching pads for the latest space probes.

Regardless of the specialty area, civil engineers share a common denominator: they are doers. They are responsible for serving their communities by improving the quality of life. Service to the community—its development and improvement—sums up what civil engineering is all about.

The Nature of the Work

Civil engineers are involved in planning, designing, and managing a wide variety of projects. Therefore, they may find themselves at a computer work station, in front of a public hearing, or on a project work site. Like other engineers, most civil engineers are part of a team that may include other engineers, scientists, contractors, project owners, architects, bankers, lawyers, and government officials.

According to the American Society of Civil Engineers, there are eight major specialties within civil engineering. These include the following areas.

Structural Engineering   Structural engineers are planners and designers of structures of all types, including bridges, dams, and power plants.

They also plan and design support for equipment, such as special structures for offshore projects, space programs, transmission facilities, and telescopes.

The structural engineer analyzes the forces that a building or structure must resist, such as its own weight, wind, temperature extremes, and natural hazards—fire, earthquake, and flood.

Taking these variables into consideration, the engineer determines a combination of appropriate building materials that can include steel, concrete, wood, and other materials.

If a structure must support a load and is made of steel, aluminum, concrete, or other materials, the structural engineer will do the planning and design to ensure the correct combination of these materials and that proper application and assembly take place. Structural engineers are team players. They normally work with architects, mechanical and electrical engineers, contractors, representatives of project owners, lawyers, public officials, and financial specialists.

Urban Planning.   Urban engineering/planning combines construction engineering and community and urban planning. Construction engineers, as urban planners are known, use both technical and management skills to plan and build public and private projects as well as commercial developments.

They apply knowledge of construction methods and equipment and principles of planning, organizing, financing, managing, and operating construction enterprises. They plan each job from beginning to end, determining the equipment, plant, and workers required to complete the task. They also estimate costs and monitor expenditures.

In urban and community planning, these civil engineers are concerned with the total development of a community. This involves:

- projecting street patterns
- identifying park and recreational areas
- determining and identifying areas for industrial, as well as residential, growth
- consulting with local authorities on the integration of the community with mass transportation and other related facilities

They also coordinate the activities of almost everyone involved in a project, so they must be as people oriented as they are technically knowledgeable.

Environmental Engineering.   Environmental engineers design and supervise systems that provide safe drinking water and prevent and control pollution in water, on land, in the air, and in the groundwater supply. Their efforts are essential to many areas of water resource management, including the design of water treatment and distribution systems, waste water collection, treatment facilities, and the containment of hazardous wastes.

As a community expands, increased demands are placed on the water and waste treatment facilities, and the environmental engineer plays an increasingly important role in providing for the orderly growth of a community as well as for its continued quality of life.

This becomes more important as people move from crowded cities to what was once rural America. With the people come industry and then jobs. This influx places an increased demand on the public works of a community, so the environmental engineer becomes as much a planner and facilitator as he or she does a traditional engineer. Through the engineer's sound management and planning, growth remains orderly as the physical plant expands to meet the growing needs of commerce and the residential community. The environmental engineer, by virtue of his or her mission, will be working closely with the engineers specializing in construction engineering and in community and urban planning.

Transportation and Pipeline Engineering. Transportation engineers are involved with the safe and efficient movement of people and goods. They design and maintain all types of transportation facilities, including highways and streets, mass transit systems, railroads and airports, and ports and harbors.

Transportation engineers apply technological knowledge and an understanding of political, economic, and social factors in their projects. They work closely with urban planners and construction engineers, since the quality of the community is directly related to the quality of its transportation system.

The transportation of gas, oil, coal, and other commodities through pipelines has created another civil engineering specialty—the pipeline engineer. This specialty combines knowledge of hydraulics, geotechnical engineering, and the structural properties of pipeline materials to ensure a steady, reliable flow of these vital commodities. Like the transportation engineer, the pipeline engineer works together with the construction engineer, environmental engineer, and urban planner.

Geotechnical Engineering. Geotechnical engineers analyze the properties of soil and rock that support and affect the behavior of structures, pavements, and underground facilities. In conjunction with environmental engineers, they evaluate the potential settlements of buildings, the stability of slopes and fields, seepage of groundwater, and the effects of earthquakes.

With structural and construction engineers, they take part in the design and construction of earth structures (dams and levees), foundations of buildings, and other construction projects such as offshore platforms, tunnels, and conventional dams.

Photogrammetry, Surveying, and Mapping. Civil engineers are also involved in making precise measurements of the earth's surface to obtain reliable information for locating and designing engineering projects.

Currently, civil engineers make use of satellites, aerial and terrestrial photomapping, and computer processing of photographic imagery. Radio and TV signals from satellites, and scanning by lasers and sonic beams are all converted to maps that give accurate measurements for use in making tunnels, building highways and dams, and plotting flood control and irrigation projects.

Engineering Management. Many of the civil engineering careers eventually lead to positions in management. In fact, some engineers—construction engineers for small projects, surveying team supervisors, and assistant municipal engineers—are able to start their engineering careers in management.

Positions in management require technical education, the ability to organize and direct workers and materials, and excellent interpersonal skills. As these skills develop, so does the amount of responsibility the civil engineer handles, until eventually he or she manages large projects that have budgets in the millions of dollars.

Interpersonal skills combined with well-developed communications and engineering skills can give any engineer a distinct advantage when seeking a management position.

Teaching Engineering. Last, but not least, many civil engineers, once they have earned advanced degrees, share their knowledge and experience as teachers or professors of engineering.

Beginning as an assistant instructor or associate professor, the engineer can progress to full professorship or head of a department, teaching both undergraduate and graduate students.

Teaching is an especially rewarding way to relay knowledge that has been acquired over the years to prospective civil engineers. In addition, because of their vast experience, faculty members are frequently asked to serve on technical boards, commissions, and other authorities associated with major engineering projects.

The Settings in Which Civil Engineers Work

In addition to engineering consulting firms, federal, state, and local governments are major employers of civil engineers. Civil engineers also work for utility companies, oil companies, telecommunication businesses, even toy and athletic equipment manufacturers. Civil engineers often work as consultants in their special area of expertise.

Unlike some other areas of engineering, civil engineering offers numerous opportunities to work outdoors. Some civil engineers spend as much as 75 to 80 percent of their time outdoors. There are also opportunities that allow civil engineers to spend most of their time indoors. The diversity of work and of work settings makes civil engineering a dynamic profession.

Training and Other Qualifications

Entrance into a civil engineering program may be at the freshman level following high school, or at the junior level after completing an approved two-year junior college program.

The typical four-year program of study in civil engineering includes one year of mathematics and basic sciences; one year of engineering science and analysis; one year of engineering theory and design; and one year that includes social sciences, humanities, communications, ethics and professionalism, along with electives that complement your overall education.

The curriculum offered by different colleges varies in detail. The exact information is listed in their college catalogues. A list of accredited civil engineering programs can be obtained from the Accreditation Board for Engineering and Technology (ABET).

Basic science: Math, physics, and chemistry.

Engineering: Engineering and scientific programming, introduction to engineering, mechanics, soil mechanics, engineering geology, strength of materials, dynamics, analysis of determinate and indeterminate structures, hydraulics, highway geometrics, surveying.

Engineering design: Engineering design, design of steel and concrete structures.

Communications: English, speech, technical writing, computer languages, and graphics.

General education: History, philosophy, psychology, sociology, anthropology, art, music, and literature.

Outlook for the Future

Since the beginning of World War II, technological change in the United States has taken place at an ever-increasing rate. This rapid pace of technical change has produce changes in the style and standard of living throughout the United States and Western Europe. Increasing emphasis and dependence on technology has produced an increase in the demand for engineers which has been relatively constant since the 1940s.

However, shifts in engineering concerns have caused periodic changes in the demand for a given engineering field. For example, in the mid 1970s a very strong demand occurred for civil engineers to work in pollution control, energy supply, and environmental protection. In the early 1980s, in contrast, a national emphasis on defense and computer technology led to a much lower demand for civil engineering graduates. What this means is that there have always been jobs for civil engineering graduates, but at times graduating civil engineers had fewer job offers from which to choose.

The most important factor to consider is that there has always been a steady demand for civil engineering graduates. There is very little likelihood that this demand will ever disappear. The types of activities in which civil engineers are engaged are critical to the continuing life of the United States. These activities range from the most fundamental, such as the supply of drinking water, to the most sophisticated, such as the construction of space stations. The decrease in spending for public works (such as transportation systems, sewer systems, and water supply systems) during the early 1980s has simply created a serious problem for coming generations, who will be required to repair and renovate these facilities.

Such repair and renovation obviously will require civil engineers. The bottom line is that employment prospects for civil engineers, which have been good to excellent in the past, are good now and will likely become even better.

The infrastructure (bridges, highways, rail systems, dams, etc.) of our country is currently a high priority of business and government. Without a strong infrastructure, the economy cannot grow and expand. This realization and need to rebuild and refurbish our roads, bridges, and buildings is creating a new demand for civil engineers.

Earnings

According to the 1992 Lindquist-Endicott Report, the average monthly starting salary for a person graduating with a BS degree in civil engineering was $2,883.00. For those graduating with a master's degree in civil engineering, the average monthly starting salary was $3,279.00.

Additional Sources of Information

American Society of Civil Engineering
Student Services
345 East 47th Street
New York, NY 10017
1-800-548-ASCE

ELECTRICAL ENGINEERING

Of all the engineering disciplines, the field of electrical and electronics engineering seems to be the single area most in touch with today's world. As a result of their knowledge of electrical phenomena and technology, electrical engineers work in a large number of capacities that support total engineering efforts in the industries that employ them.

According to the *Dictionary of Occupational Titles,* electrical engineers apply the laws of electrical energy and the principles of engineering to the generation, transmission, and use of electricity. Electrical engineers design everything from power generating systems in dams to tiny electronic circuits for spacecraft. They create the electronic components that run computers, televisions, stereo systems, and automated factories.

The Nature of the Work

There are numerous subclassifications of electrical engineering that encompass virtually every facet of our lives. Little, if anything, we take for granted in our daily existence would have been possible without the intervention of skilled electrical engineers. In fact, the actual classifications and categories rapidly blur in today's technology. It is very difficult to distinguish where an electronics engineer's contributions end and a specialist-engineer's in control systems begin, yet both professionals are electrical engineers. In fact, the term *electrical engineering* is really very limiting, as there are more than 30 classifications under the umbrella of electrical engineering. However, there are four well-recognized branches of electrical engineering—power, communications, electronics and control systems.

Power.   Electrical engineers who specialize or concentrate in the power field are involved in power *generation, transmission, distribution, application,* or in combinations of these branches. The very broad field of manufacturing and machine design and the field's applications are closely related to power engineering.

Power generation.   Power generation is merely converting energy from a static form to one that is adaptable to our needs. Engineers working in this field must design systems that can utilize static forms of energy (water power, solar power, fossil fuels, and chemical agents) to produce usable electric power.

Ideally, engineers strive for the maximum in efficiency; however, there is no such thing as totally efficient power. Determining how this energy is to be converted is the responsibility of power generation engineers, specially trained electrical engineers who design the best methods of conversion of one form of energy (static) to the desired form of energy— electrical power. Because there are different sources of power, the techniques which electrical engineers use to generate or design power generation vary.

Water power is harnessed through the use of hydraulics. It is used to turn electric generators called turbines, which convert the water to motion, which in turn spins the turbine blades and drives the generators. The end result is electric power.

Fossil fuels—oil, gas, and coal—are nonrenewable resources with particular problems of their own. Besides the problem of a potentially uncontrolled demand on an inherently limited supply, there is the overriding problem of the environmental pollution that is produced when fossil fuels are converted into electrical power. Specialized electrical engineers deal with both aspects of the problem: how best to utilize a limited resource to produce power and how to generate this power with the least amount of environmental damage (pollution). These design criteria apply from the basic drawing board stages to the final operation of the power plant.

Nuclear, geothermal, and solar power generation offer some of the greatest challenges to electrical engineers. These power sources are considered almost limitless; however, harnessing and converting them to usable power can be quite challenging.

Geothermal energy conversion, or the generation of electrical power from natural sources of heat deep within the earth's crust, is experimental. Electrical engineers have coupled conventional steam-driven, turbine-powered generation with some radically new technology, enabling them to convert the earth's core heat into a reliable and controlled source of steam power. This power will drive tomorrow's electrical generation plants. Though nonrenewable, geothermal energy can probably, according to the experts, be considered limitless, as a fairly large percentage of thermal energy is returned to its source.

Solar power generation, although used extensively in our space program, has yet to achieve a cost-effective place in the engineer's armory of power generation schemes. Converting energy from the sun to usable quantities of electric power is still virtually in the experimental stages.

The future holds great promise for solar energy, and electrical engineers are working on more cost-effective methods of converting our limitless supplies of sunlight and heat from geothermal sources to electricity to meet our future needs.

Nuclear power, once thought to be the answer to our power requirements, produces nearly as many problems for the engineer. While nuclear power is efficient, it has inherent dangers. Under carefully controlled applications, engineers have incorporated both efficiency and safety into the design and operation of a nuclear power plant, which converts nuclear energy into heat, which in turn produces steam from water. The steam then drives turbines and produces electricity.

Compared to other forms of power generation, nuclear power is clean and relatively pollutant free. However, the way the power is produced, or the conversion method, makes for some very unique and interesting problems for the electrical engineer. Security—the physical containment of the nuclear power source—is of utmost importance. Likewise, there must be stringent monitoring of the source and means of ensuring that the potentially hazardous energy is kept in its place, under control. Special techniques and controls must be developed to prevent and guard against leaks and possible equipment failures.

In addition, the electrical engineers designing these projects must take into account and plan for acceptable means of disposing waste generated by nuclear power. These factors make for challenges that require the expertise of several electrical engineering classifications.

Transmission and distribution. While electrical engineers devise efficient ways of making power, others work at maximizing transmission and distribution efficiency.

The transmission and distribution of electrical power is governed by strict rules of physics. There are losses no matter how well a distribution and transmission network is designed. In order to minimize these losses and provide the maximum power transfer, electrical engineers design and implement schemes that utilize power transformers to convert raw electrical power (from the source) to a high voltage for more effective transfer over long distances with less loss.

Applications. As power reaches its desired location, other electrical engineers have already been at work developing effective methods of using the power. These engineers specialize in power applications that can range from the design of lighting systems to schemes for the electrical motors driving some of the very latest mass transportation systems.

These engineers must juggle several problems at the same time. They must be conscious of the overall cost impact that technological developments will have on the end product and, at the same time, be energy conservationists, squeezing every last possible watt of energy out of a power source. These same engineers also develop the machines that manufacture the products that make our life-style what it is. As such, they work in automotive engineering and paper and steel manufacturing as well as a host of other large and small consumer-related businesses.

Communication.   Electrical engineers who specialize or concentrate in the communication field are involved in *equipment engineering, transmission engineering, switching and circuits, systems engineering, traffic engineering, commercial engineering, plant engineering, or acoustical engineering.* Communications engineering has a direct impact on the production or operation of almost everything that touches our lives.

Electrical engineers in this branch design systems that receive, transmit, and deliver information in audio as well as video form. The radio, television, and telephone are all the products of electrical engineers specializing in communications. Recent technological advances, especially the joining of computer technology with information processing and distribution, have provided opportunities that were, a few years ago, only dreams to communications engineers. There are several kinds of electrical engineers directly involved in the communications process.

Equipment engineers. Equipment, or apparatus, engineers are electrical engineers involved in the design and implementation of devices that take information and translate or convert it into a form suitable for transmission to distant locations. In essence, electrical engineers build the radio and TV transmitters and receivers that provide our society with communications, whatever the form. These forms of communication include the telegraph, the telephone, and computer-assisted technology that not only processes information but transmits it to a distant location where it can be used by others. These link-ups also include aerospace engineering, in which satellites serve as both transmitters and receivers and so act as relay centers high above the earth, linking distant points as if they were next door.

Transmission engineers. Transmission engineers provide the pathways or channels for communications signals (which can be very weak), amplifying them as necessary and working out various details to make transmission as reliable as possible. An understanding of the study and science of wave propagation—the effects the earth and the atmosphere will have on a radio signal—is very important to the transmission engineer's daily activities. These engineers are responsible for the accuracy and quality of these radio and TV signals as well.

Switching and circuit engineers. These engineers specialize in switching circuitry. They are the control, direction, and "glue" for the entire communications effort. These engineers make it possible for us to directly dial telephone numbers in cities and countries far from our homes, without the assistance of an operator.

In addition, these engineers design and develop the almost "human" switching centers. These centers constantly monitor the use and quality of communications traffic and transmission and, if they sense a fault, re-route the communications with hardly a flicker or lost bit. Switching and circuit engineers' tools include circuits, components, batteries, exotic power supplies, and banks of computers with programmed responses to changing conditions.

Systems engineers. Systems engineers specialize in improving the overall performance of switching systems. They are customer oriented, and they improve customer service by introducing new features and reducing communications costs. The excellent performance of the American telephone system demonstrates just how effective the systems engineer is in today's communications world.

Traffic engineers. The traffic engineer, yet another member of the communications team, is the direct link between a communications system and its users. Traffic engineers are concerned with the availability of adequate service facilities to handle not only normal system traffic but overloads as well. Their work is a combination of accounting, engineering, and planning. They study equipment capabilities and how to plot these capabilities against customer patterns of use. They study circuit operating efficiencies as well. Thus they ensure that a system can handle any demand without having excessive or unnecessary idle circuit time. They can be considered the "auditors" of the communications engineering field.

Commercial engineers. Commercial engineers specialize in the service aspects of communications. The public's needs and its reaction to services, costs, and other limitations are among their concerns. These engineers are also responsible for balancing rates (what a service costs the communications supplier) with revenues (how much the supplier can charge the user of this service).

Plant engineers. Plant engineers deal with plant costs and extensions. They are also concerned with planning for the expansion of existing communications facilities. Some of their duties are not usually associated with electrical or communications engineers. These duties include general day-to-day building operations and negotiating with local communities to secure land and rights of way.

Acoustical engineers. While the other members of the communications team handle substantial portions of the total effort, their work would be lost if it were not for the acoustical engineer. Acoustical engineers specialize in the design and implementation of devices that convert sound to a form suitable for transmission over radio waves and then reproduce it through loudspeakers. They are also concerned with the design of studios, halls, and other public facilities where people go to hear and see movies, concerts, and other forms of entertainment.

Acoustical engineers are also concerned with sound levels and noise pollution. Without them, what might be music to one person could be a source of pain to others. They constantly monitor the levels of sounds and formulate charts identifying what is a safe sound and when a sound becomes hazardous to people. Many of these engineers are also employed in industry to help ensure that workers are not subjected to dangerous noise levels in offices and industrial sites.

Electronics. Of the four well-recognized classifications of electrical engineering, perhaps electronics has the distinction of being the most visible and, therefore, the most "glamorous." There are many subcategories of electronics engineering, ranging from the mix of electronics and physiology (biomedical or clinical engineering) to computers and data processing. Along with these fields there are literally dozens of fields that materially and directly affect our lives and our futures, such as consumer and home electronics, including televisions, radios, CDs, and VCRs—all products of electronics engineers.

Computer systems, in shrinking from room-sized to laptop, have opened new, challenging opportunities to electronics engineers. The development and enhancement of electronic aids to navigation have made travel safer and opened new opportunities for the electronics engineer, also.

Biomedical or medical electronics is another growth opportunity for the electronics engineer as well as bionic replacements for body parts; artificial hearts; pacemakers (electronic regulators for failing hearts); and devices based on sonar that help the blind "see."

Control Systems. This specialty deals with the analysis and design of automatic regulators, guidance systems, numerical control of machines, computer control of industrial processes, and robotics. Electrical engineers in this area are concerned with the identification of system stability, system performance criteria, and optimization.

Control systems are essential in the automation of complex manufacturing processes used in making products such as gasoline, detergent, appliances, food and medicine, and household items that are used every day. These engineers design the devices that manufacture cars, cut out patterns, assemble parts, move objects, and control our environment.

Very few home appliances lack a control system. Precision, required in the manufacture of many electrical, electronic, and mechanical products, is made possible by the control systems engineer. Together with other electronics engineers, they also produce machines to make other machines—they produce robots.

Related Fields

Many individuals trained as electrical engineers apply their knowledge to related fields. A few of the common related applications are discussed here.

Electromechanical.   Probably the most common merger of electricity with another field occurs when mechanical design is required to activate some new electrical device. Operation of the machine may depend upon some intricate mechanical apparatus without which the innovative electrical design is useless.

A good example of the interdependence of electrical and mechanical design is the anti-aircraft gun director developed during World War II. Plane search radar was used to locate enemy aircraft and to take continuous bearings in horizontal and vertical planes. These bearings were fed to an electronic computer which calculated the course of the plane and predicted its location at the time a shell could be sent to intercept it. The time and location were dependent on plane speed and course and on the trajectory and flight time of the shell. As soon as the plane could be tracked by telescope more accurately than by radar, the controls of the computer were shifted to optical observers. This was more intricate than most electromechanical projects, but it illustrates the interdependence of electrical and mechanical design.

Electrochemical.   Opportunities for electrical engineers in the chemical and allied industries occur primarily in the power field.

The chemical composition of materials used in electrical applications may be critical in some applications. To ensure consistent results, the electrical engineer may have to delve deeply into the chemical composition of matter. The presence of minute impurities in lead plates for storage batteries for many years led to erratic performance. It was only through careful study and experimentation that unwanted impurities were weeded out and allowable amounts of antimony, which improved performance, were introduced.

Industry. While some schools give courses leading to a degree in industrial engineering, the more common approach is for a practicing engineer in one of the major fields to swing gradually from his or her chosen specialty to more general industrial practice. In fact, such a move is the logical outgrowth of an engineer's success in an industrial enterprise, just as the outstanding power engineer may be selected for management in a public utility. There may be this difference, however: whereas the utility executive is largely concerned with power and is head of an organization composed mostly of engineers, the industrial engineer is apt to grow away from the electrical field into the broader field of industry. While he or she may have a small staff of engineers, his or her principal activities will be in the management of a production organization.

Heating, Air Conditioning, Refrigeration. You may be surprised to learn that heating, air conditioning, and refrigeration can be considered branches of electrical engineering. Usually, they are associated with the mechanical field. Many electrical engineers have become interested in the electrical specialties involved in these disciplines and have gradually grown to embrace all phases of the subject. Actually, these are three fairly well-defined fields which are very closely related and which can be combined to excellent advantage.

Heating and ventilation are coming more to involve electric appliances for regulation, controls, and circulation. Additionally, electrical engineers are much interested in the possibility of solar heat installations and of storage of heat for equalizing winter and summer temperatures. Some very interesting experiments have been conducted along these lines. The applications employ the same techniques as refrigeration. Hence the combination of refrigeration with heating, ventilation, and air conditioning comes as a matter of course.

Sales. Many organizations with a product to sell are glad to employ electrical engineers because they have the background to discuss related electrical problems with prospective customers. As a general rule, sales ability is relatively well rewarded, and most engineers who enter this field remain in it, even though the application of engineering in their work is slight.

Another sales opportunity for the engineer is as a specialist who combines engineering and selling. For example, the manufacturer of arc welding equipment may require as a salesperson an engineer who can size up the requirements of a prospective customer, design welding equipment adapted for the customer's plant operation, and so place the order in such a way that both supplier and customer will be pleased. He or she must further be able to supervise and test the installation and even make corrections where necessary. This type of sales ability is required in many lines of electrical equipment.

Public regulation. Of a somewhat similar nature is the engineering service rendered to public regulating commissions. This work consists largely of valuations of plants, depreciation studies, and determination of rates. Here, again, experience is important although not essential.

The Settings in Which Electrical Engineers Work

Consulting. Engineers who become consultants usually have special expertise in a specific engineering discipline and have practiced successfully in that field for many years. Some consultants are retired engineers who wish to remain active professionally; others are practicing engineers who are employed by consulting firms.

The consulting engineer may specialize in appraisals or rates. He or she may serve as an expert witness in public hearings or in litigation. Several of the larger consulting organizations are prepared to render complete service, including preliminary surveys, financing, construction, operation, and management on large projects. One phase of consulting engineering that has proved attractive is the development of specialty products, starting with the client's idea and including the research, patent protection, manufacturing, and marketing. On a profit-sharing basis, the consulting engineer may be rewarded handsomely on such a project.

Government Service. Government service is attracting its quota of engineers at federal, state, and local levels. Openings for electrical engineers in the federal service are provided by the armed forces, the Bureau of Standards, and various commissions. The Air Force and Signal Corps, particularly, have many engineering positions in communications, in navigational aids, and in guided missiles.

Industry. Almost every part of the industrialized world's economy employs electrical engineers. In addition to consulting firms, government agencies, and universities, the following is a partial list of the industrial settings in which electrical engineers can be employed:

Aeronautical/Aerospace
Automotive
Chemical and Petrochemical
Computers
Construction
Defense
Electric Utilities
Electronics
Environmental
Food and Beverage

Glass, Ceramics, and Metals
Machine Tools
Mining and Metallurgy
Nuclear
Oceanography
Pulp and Paper
Textiles
Transportation
Water and Wastewater

Teaching.   Many engineers go directly into teaching or turn to it after years of successful practice. The profession is rewarding in the influence it allows engineers to exercise and in the contacts it offers. Teaching usually affords engineers leisure time which may be put to good use in writing or in consulting. Not infrequently, research undertaken in an engineering school leads to worthwhile inventions. In recognition of this, some schools have set up foundations to reward research that develops ideas and products for the benefit of society.

Training and Other Qualifications

In addition to the basic science and math that all engineering students take, electrical engineering students take required courses in mathematical logic and set theory, algorithms, numerical methods and analysis, probability and statistics, and operating systems. They also take courses in computer science and programming techniques. Depending on the expertise of faculty members at a particular university, students then specialize in one of the branches or subdivisions of electrical engineering: power generation, control systems, communications, or electronics. Each of the professional societies in electrical engineering can provide a list of institutions offering academic programs beyond high school in electrical engineering fields.

Outlook for the Future

The phenomenal growth of the electronics industry has fueled a growing demand for electrical and electronics engineers and has led to an increased interest on the part of young engineers in this field. The rapid production gains in the electrical equipment industries continue to lead to new job growth in this area of engineering.

Earnings

According to the 1992 Lindquist-Endicott Report, the average monthly starting salary for a person graduating with a BS degree in electrical engineering was $2,892.00. For those graduating with a master's degree in electrical engineering, the average monthly starting salary was $3,420.00. New Ph.D.'s in electrical engineering reported an average monthly starting salary of $4,876.00.

Additional Sources of Information

Institute of Electrical and Electronics Engineers
345 East 47th Street
New York, NY 10017

Instrumentation Society of America
67 Alexander Drive
P.O. Box 12277
Research Triangle Park, NC 27709

Society of Photo-Optical Instrumentation Engineers
P.O. Box 10
Bellingham, WA 98227-0010

INDUSTRIAL ENGINEERING

Industrial engineers have come a long way from the days when they carried clipboards and stopwatches. Now, industrial engineers apply their knowledge to a wider range of issues and in a broader array of settings. As a result they require the broadest possible exposure to a variety of engineering disciplines.

Increasingly, industrial engineers serve as the link between engineering and management. In fact, one chief executive officer has referred to industrial engineering as "the stepping stone to management."

The primary goal of an industrial engineer is to improve productivity. Therefore, recent changes in the way goods are produced or built have dictated changes in the industrial engineer's duties.

The Nature of the Work

Industrial engineers have a strong desire to serve human needs and enjoy working with people. They tend to take the "big picture" approach. They plan, organize, and carry out projects in a wide variety of settings by balancing the needs and abilities of people with the availability and characteristics of materials and energy as well as equipment and facilities. They seek the best alternatives to bridge the gap between management and operations. With this broad focus industrial engineering has many facets.

According to the Institute for Industrial Engineering, industrial engineers become involved with such things as: advancing manufacturing methods utilizing robotics; computer and information systems; energy management; engineering economy (a financial area); facilities planning and design, including material handling; human factors or ergonomics; human resources management; operations research and computer simulation; organization and job design; production and inventory control; quality assurance; as well as warehousing and distribution work measurement.

Industrial engineers get involved with such things as:

Long-range planning and facilities design for a major transportation facility.

Robotics programs at a major automotive manufacturer.

Assisting in the design and installation of operations systems for semi-conductor facilities.

Creating more productive work flow within hospital and other health institutions.

Designing a computer-based management information system for any organization.

Industrial engineers are concerned with performance measures and standards, research of new products and product applications, ways to improve use of scarce resources, and many other problem-solving adventures. Industrial engineers relate to the total picture of productivity improvement where productivity means getting the most out for the least put in.

Industrial engineers also look at the right combination of human resources, natural resources, and manmade structures and equipment to optimize productivity. They address the issue of motivating people as well as determining what tools should be used and how they should be used.

Industrial engineers are involved in such areas as: operations research, applied behavioral science, and systems engineering.

Operations Research.   In this area, industrial engineers describe a situation in mathematical models to determine the best course of action to recommend.

Applied Behavioral Science.   This area combines engineering principle with behavioral sciences such as sociology, psychology, and anthropology to improve the management function. Industrial engineers study how organizations work and how they can work better. Their approach is scientific and quantitative. The management of technology is a major application of this area.

System Engineering.   Industrial engineers in this area are concerned with complex systems in manufacturing engineering, production, transportation, housing, health care delivery, energy allocation, environmental control, criminal justice, and education.

The Settings in Which Industrial Engineers Work

Industrial engineers can pursue their careers in a wide variety of work settings. Industrial engineering is performed in all major sections of industry as classified by the federal government's Standard Industrial Classification (SIC) code. In addition to the manufacturing sector, industrial engineers are employed in such diverse areas as accounting, merchandising, banks, hospitals, government and social service agencies, transportation and construction industries.

According to the College Placement Council, the most aggressive new recruiter of industrial engineers is the electrical and electronics machinery sector.

Training and Other Qualifications

Mathematics and science play a key role in the industrial engineer's knowledge. At least three years of high school math, including calculus, and courses in chemistry and physics are important preparation for any engineering career.

At the college level, industrial engineering students take courses in engineering economics, person-machines engineering, devices organizational development, and system evaluation. Computer simulation is an especially effective tool for systems analysis. It is also recommended that industrial engineers study human factors, including courses in psychology, biology, and the social sciences.

Industrial engineering is offered at approximately 100 accredited universities in the United States and Canada. A list of these institutions can be obtained from the Institute of Industrial Engineers.

Outlook for the Future

Like most of the engineering professionals, industrial engineers are in great demand not merely because they are engineers but also because of how they apply their engineering skills.

The increased demand for industrial engineers is due in great part to the need for organizations to raise the level of their productivity through careful, systematic approaches. Any profit-making organization must have a high degree of productivity to compete in both a local and a global marketplace. Even a nonprofit organization uses industrial engineers in order to be able to sustain its position as a useful service entity.

Because of the pointed demand for industrial engineers, this profession is extremely attractive in terms of financial rewards. Business is predisposed to reward those who make it more profitable, and the industrial engineer does exactly that. In fact, salaries for industrial engineers are among the highest for all engineering disciplines. Many industrial engineers move quickly into management, making the outlook for continued growth excellent. Increased competition from industries abroad will also probably sustain a growing need for industrial engineers.

Earnings

Beginning salaries range in the top group of high-paying engineering disciplines.

According to the 1992 Lindquist-Endicott Report, the average monthly starting salary for a person graduating with a BS degree in industrial engineering was $2,790.00. For those graduating with a master's degree in industrial engineering, the average monthly starting salary was $3,187.00

Additional Sources of Information

Institute of Industrial Engineers
25 Technology Park
Norcross, Georgia 30092

MATERIALS SCIENCE AND ENGINEERING

In many ways materials science and engineering, including metallurgical and ceramic engineering, plays a key role in the development of new technologies. Materials scientists and engineers deal with producing materials that have properties that make them suitable for practical use.

The Nature of the Work

Materials scientists and engineers study the properties of various types of material (i.e., ceramics, polymers (plastics), metals, electronics materials). They also study how materials are made and how they behave under different conditions. Some materials scientists and engineers design materials such as alloys and composites for special uses in the space shuttle and other high-tech industrial areas.

Metals. Metallurgy is the study of metals. It is a very old field, having its roots thousands of years ago during the Bronze and Iron ages, when cave dwellers first began to make use of metals. Metallurgy has developed from the early attempts of the cave dwellers to today's production of lightweight alloys (combinations of metals), which exhibit special properties.

Today, newer generations of engineered metals—combinations of existing metals—together with advances in engineering produce new materials with properties that meet or exceed those of the more traditional metals.

Metallurgical engineers work at extracting metals from ores and refining, alloying, casting, fabricating, and heat-treating them to develop better methods and techniques to meet our metals requirements. This work is divided into three categories:

1. Process metallurgy
2. Physical metallurgy
3. Materials science

Process metallurgy is part of plant design and processing. The engineer specializing in this facet of the materials engineering field will have taken many of the core courses required for other engineering fields such as chemical, mechanical, and electrical engineering.

Physical metallurgy is the study and analysis of the structures of metals as they relate to their physical properties.

Materials science combines several principles of physical metallurgy, ceramics, and polymer chemistry. It emphasizes the study of the properties and uses of metals, ceramics, polymers, and engineered composite materials. The materials science specialist integrates most, if not all, of the techniques of materials engineering to produce new materials or to make existing materials more useful as the needs of society and industry change.

Ceramics. Ceramic materials were first used about eight thousand years ago by people living in what is now known as Turkey. They made dishes and pots from clay and then fired them to make the clay hard and smooth. These potters and dish makers could be considered the first ceramic engineers.

Today, ceramic materials are used for a variety of purposes. The space program is a large consumer of ceramics. The heat-shield tiles on the space shuttle, as well as space capsules and even missile nose cones, are made of specially formulated ceramic materials. Closer to home, the cements and bricks used in the construction of houses are also ceramic.

Ceramics are nonmetallic, inorganic materials produced from raw materials that are, for the most part, abundant and relatively inexpensive. Some of the materials are chemically synthesized and are not available in a natural state.

Ceramic engineers work with materials having a wide range of characteristics that can be exploited in the development of new products. For example, ceramic materials that are insulating yet magnetic make the household microwave oven possible. Replacements for human bones and teeth that are durable, lightweight, and strong are also made possible by ceramic engineering.

Glass. Ceramic engineers are producing new applications for existing products such as glass. Glass fibers are now replacing metal wires in communications systems. Telephone companies can transmit voices and data using laser technology and optical glass fibers. A major advantage is that more information can be sent through a relatively smaller cable.

Additionally, fiber optics—as the application of glass fibers in communications is known—offers virtually no interference and less resistance than is commonly associated with conventional metal cables. For this reason, many existing telephone lines are being replaced with cables made of glass fibers. This technological advance would not have been possible without the efforts of the ceramic engineer.

Electronics. Two segments of the electronics industry rely heavily on the skills of the ceramic engineer:

1. Electrical utility companies
2. Semiconductor manufacturers

Electric utilities require huge ceramic insulators on their high-voltage power transmission lines. They also use smaller insulators on the poles in front of your house or apartment. These insulators protect workers and the public from stray electrical voltages and also prevent the loss of electricity to the ground.

The semiconductor industry could not exist without ceramic engineering. Ceramics are used as insulators and building blocks for the integrated circuits, or chips, that have made so many of today's products possible and affordable. These chips are used in calculators, watches, stereos, televisions, and communications satellites. There are other uses and applications for integrated circuits and semiconductors, but it would be safe to say that without the ceramic engineer's contributions, we wouldn't have the standard of living or the electrical and electronic aids we take for granted.

Other electronic components also rely on ceramic engineering. These include capacitors, resistors, and modern sensors that convert information to electrical impulses for further processing and are used in control systems as well as in medical electronics technology.

Plastics.   Materials engineering is also the field that works with and develops new applications for plastics and other *polymers*, or materials produced by combining chains of hydrocarbon molecules, themselves called polymers.

How the hydrogen is combined with the carbon and how these chains are strung together are challenges for materials engineers. New plastics have replaced more traditional materials, such as metal, glass, and wood, in a wide variety of applications, such as in automobiles, computers, furniture, and packaging. Engineered plastics could very well revolutionize the world of materials, and the materials engineer is at the forefront of this development.

Composites.   Composites, as the name implies, are combinations of materials. Often, thin fibers of metals or nonmetals are literally woven into a fabric. This fabric is placed in a mold and covered with an engineered plastic resin. The result is a lightweight, strong, and durable material that combines properties of the base materials with corrosion resistance and flexibility. Often, these engineered materials are clearly superior to the separate materials that compose them.

The aircraft industry, the space program, and even the auto industry are making increased use of these engineered materials. The stealth technology, by which an object is made invisible to radar, would not be possible without composite technology. In this application, materials are deliberately selected that will absorb rather than reflect radar waves, rendering the object mostly invisible to the radar operator.

In other, nonmilitary applications, substituting composite materials for more conventional metals or alloys has allowed private corporate jet aircraft to fly farther, and faster, with less fuel, than the noncomposite aircraft of just a few years ago.

Automobile manufacturers are major users of composites. The composite materials are used to replace metals that are too heavy, not corrosion-resistant, or not as strong. Lightening the auto's weight by replacing heavy metals increases its fuel efficiency while ensuring corrosion resistance and strength.

Boats, campers, and trailers are made of fiberglass, a composite material that is much more useful and durable than the traditional wood or metal products it replaced. Fiberglass is a good example of composite engineering brought to the consumer level.

Materials engineers are employed in a variety of situations, including the following areas.

Research. Working with the building blocks of matter, engineers can unlock the secrets of nature. Basic knowledge is discovered that can benefit people everywhere.

Extractive engineering. This involves supplying materials out of their natural state as well as extracting them from recycled products. Recycling permits reuse of many materials that are nonrenewable and restores useful materials to our ever-diminishing inventory of natural resources.

Process engineering. This field deals with the production of high-quality, reliable, uniform, predictable materials. The materials engineer is a vital contributor to this effort.

Applications engineering. Applications engineers develop new ways, new processes, and new materials to make virtually any product. The materials engineer helps a company by applying technology to improve existing products or to produce new, better ones.

Management. Engineers, because of their systematic approach to problem solving, have a clear understanding of different aspects of problems. For this reason, many serve as managers and supervisors, using their investigative skills to identify and solve a broad variety of problems, including the allocation of resources—both material and human.

Sales engineering. The engineer's skills in matching materials to products and products to applications combined with his or her communications skills make for the best of all possible sales representation. Thus, a materials engineer can succeed where others might fail in the sale and marketing of products.

Service engineering. Materials engineers apply their problem-solving techniques and communications skills to help customers solve problems they have with products the engineer may have developed.

Consulting. As an independent materials engineer, you could serve a variety of clients who have equally diverse needs. Many companies require the skills that only a materials engineer possesses and, unless they employ materials engineers, must hire independent ones as consultants. The consultant frequently provides a smaller company with the competitive and technological edge it requires in order to grow.

Writing and teaching. Both build on communications skills and a desire to impart information. Materials engineers can promote technology and train tomorrow's problem solvers by teaching or publishing about their field and experiences. The ability to take a complex process, solution, or technology and explain it so that students can benefit is the mark of a true engineering professional.

The Settings in Which Materials Scientists and Engineers Work

Materials engineers will face many challenges as they meet today's needs and facilitate tomorrow's technology. These challenges and opportunities are in areas such as research, food, discovery, development, energy, conservation, production, and pollution.

These complex problems and challenges are global in scope and will require the cooperation of industry, government, and all nations. Materials engineers will be at the head of the problem-solving team, engineering new materials to replace older, nonrenewable resources and constantly improving the quality of life for all the world.

In order to accomplish these tasks, they will be employed in the following areas

Materials producing companies. These produce better materials more efficiently and cleanly and provide the raw ingredients to make the advanced machinery and equipment needed to solve other technological problems.

Manufacturing companies. These utilize the services of the materials engineer to more effectively manufacture products such as cars, appliances, electronics, aerospace equipment, other machinery, and medicine. The materials engineer plays a vital role in improving materials, processes, product reliability and safety, chemical processing, paper, plastics, and textiles.

Service companies. All companies that serve the public's needs rely on materials engineers to maintain safe, reliable service. Examples of such service include airlines, railroads, and utilities.

Consulting firms. These provide companies, institutions, and the government with independent, outside help identifying problems in materials processing and performance. They also provide guidance in developing practical, economical solutions to a variety of problems.

The government. It is a consumer, promoter, and regulator of materials, products, and technology. It needs the materials engineer or scientist to provide a flow of accurate information so that policy decisions can be based on facts, not political whims.

Research institutes. These may work under contract to the government or private industry to probe materials, processes, and product development, ensuring that when tomorrow's products are needed, the technology will be in place to produce them.

Schools and universities. These provide the materials engineer with the opportunity to share knowledge and to help train those engineers who will become the problem solvers of tomorrow.

Publishers. Like universities, these offer the materials engineer the opportunity to influence and train tomorrow's problem solvers. Using highly developed communications skills, some materials engineers write and edit scientific and engineering books, training programs, video scripts, and articles for magazines and technical journals. Information sharing and communications are essential to the continued process of problem solving.

Training and Other Qualifications

Specializations and concentrations at the undergraduate level include materials, metals, minerals, ceramics, and polymers. Many materials scientists and engineers continue their studies and obtain master's and/or doctoral degrees. The master's degree usually can be earned within two years after the BS. The doctoral degree usually involves four years.

Because so much of materials engineering occurs in the laboratory, advanced degrees are much in demand. According to the Engineering Manpower Commission in Washington, DC, there were almost as many materials engineers earning master's degrees in 1991 (690) as there were bachelor's degrees (856). And the Ph.D.'s were not far behind with 414 degrees. Few other engineering disciplines show such an even balance between undergraduate and graduate education.

Depending on career goals some materials scientists and engineers pursue study in such professional areas as business administration, medicine, management, and law.

Outlook for the Future

By understanding and improving the behavior(s) of materials, the materials scientist and engineer is much in demand by other engineers. Related engineering fields such as chemical, civil, mechanical, and electrical rely on the expertise of this small but growing pool of engineers. Discoveries in the area of superconductors and buckminsterfullerenes mean that entirely new industries will be created requiring the knowledge and skills of materials engineers. Their involvement in these efforts means that the future demand will be excellent.

Earnings

According to the 1992 Lindquist-Endicott Report, the average monthly starting salary for a person graduating with a BS degree in materials science and engineering was $2,893.00. For those graduating with a master's degree in materials science and engineering, the average monthly starting salary was $3,255.00, and for those graduating with a Ph.D., the average monthly starting salary was $4,425.00.

Additional Sources of Information

ASM International
Metals Park, OH 44073

National Institute of Ceramic Engineers
65 Ceramic Drive
Columbus, OH 43214

The Metallurgical Society of AIME
420 Commonwealth Drive
Warrendale, PA 15086

American Ceramic Society
757 Brooks Edge Plaza Drive
Westerville, OH 43081

The Minerals, Metal & Materials Society
420 Commonwealth Drive
Warrendale, PA 15086
(412) 776-9011

Society of Plastics Engineers
14 Fairfield Drive
Brookfield, CT 06804-0403

MECHANICAL ENGINEERING

Mechanical engineering is one of the most exciting engineering fields because it offers breadth, flexibility, and individuality. Mechanical engineering is a creative profession. It takes a broad outlook when solving complex problems. Mechanical engineers work in such areas as power generation, energy conversion, machine design, manufacturing and automation, and the control of engineering systems.

Mechanical engineers hold a unique position in the engineering field because they design, develop, and produce many of the tools required by other engineers. Therefore, their role expands to keep pace with technology. In the future, mechanical engineers will be vital to the success of newly emerging fields of engineering.

The Nature of the Work

Mechanical engineering is organized into three general areas: *energy, manufacturing,* and *engineering design mechanics.* Mechanical engineers are concerned with:

- The use of energy from natural sources and its economical conversion into other useful energy.
- The design and fabrication of machines to lighten the burden of human work.

- Processing materials into products that are useful to people.
- Creative planning, development, and operation of systems for using energy resources and machines.
- The education and training of specialists, frequently called technicians, to deal with mechanical systems.
- Acting as an interface between society and technology.

The American Society of Mechanical Engineers has 24 technical subdivisions. These subdivisions demonstrate the breadth of areas in which mechanical engineers work.

Air Pollution Control
Applied Mechanics
Automatic Control
Biomechanical and Human Factors
Design Engineering
Diesel and Gas Engine Power
Energetics
Fluids Engineering
Fuels
Gas Turbine
Heat Transfer
Incineration
Lubrication
Materials Handling
Metals Engineering
Nuclear Engineering
Petroleum Engineering
Power Plant Engineering
Pressure Vessel and Piping
Railroad, Aviation, and Space
Rubber and Plastics
Solar Energy Application
Textile Engineering
Underwater Technology

Figure 2.1 shows a listing of the major components of mechanical engineering activities. This broad spectrum includes most of the types of work engaged in by engineers after graduation. In actual life, the graduate may shift from one activity to another. For example, a graduate may start in the production area, then be shifted to the design area, later to the testing department, and then, in some cases, to technical sales. In

Fig. 2.1 Spectrum of activities engaged in by mechanical engineers

small companies, one individual may handle several aspects of the business at the same time, as a designer, a production supervisor, and a testing engineer—all from the same desk.

As the spectrum implies, different talents and interests are required by different departments of the same company. For example, at the left end of the spectrum we find engineers who are science oriented, technically competent, and mathematically gifted. If those engineers consult or teach, they must also be able to deal with people and have skill in communicating ideas.

At the right-hand side of the spectrum, we find engineers who deal with materials, business matters, and people. Those involved as manufacturers' representatives or sales engineers or with legal aspects of engineering and business may not require the technical competence of a research engineer but should be knowledgeable about business, accounting, economics, and people.

In a spectrum of this sort, there is no higher or lower order. The left side is no more preferred than the right side.

Research. This is the first step in solving a problem. The engineer will obtain data, devise new methods of calculation, and so acquire new knowledge.

Development. The engineer takes the information and knowledge gained from the research and begins to expand it. At this stage, a simulation or experimental device might be produced and further extended into either a process or system that approximates a solution to the problem and fits the final need.

Design. In this phase, the engineer actually conceptualizes the machine, approach, system, or combinations that will solve the problem. Careful documentation of all details is necessary to bring the solution from the mind to reality. The solution is described quantitatively and put into equation or drawing form.

Testing. To perform a test, the engineer will utilize either experimental devices or full-scale completed machines, systems, or equipment. These devices will be operated to determine performance. At the same time, the mechanical engineer is checking to determine how much use and abuse the device can withstand, its relative strengths and weaknesses, and how to improve its performance. In essence, this phase of the mechanical engineer's work is to determine that whatever is being tested will perform as it is intended and can function in the environmental conditions that were anticipated in the design criteria.

Manufacturing. Here the mechanical engineer must answer a series of questions. How is the product best manufactured? What is the most economical way of making it? What processes will be required? What are the skills and personnel needed to produce it? These questions are answered by a production engineer. The production engineer is the person who selects the equipment and machines and supervises arrangement and operation in detail. This engineer is also responsible for efficient, economical, and safe manufacturing.

Operation and maintenance. Some equipment or systems require specialized knowledge and expertise above the level technicians usually have. These systems and equipment require the continual care only a mechanical engineer can provide. These duties might be mandated by law, as are the supervision and maintenance responsibilities for a nuclear or fossil fuel plant. Federal and state regulations require a mechanical engineer at such plants to perform certain specific tasks.

Marketing and sales. When a firm offers a complex product or system, it can't rely only on a salesperson to present it to prospective clients. This is especially true in the case of systems and equipment that require formal technical backgrounds in order to understand and exploit them. In this case, the mechanical engineer would function as a sales or marketing engineer, relying on his or her technical background and communications skills to demonstrate a product or a system to a customer. Often the engineer will work with the customer to modify the basic system or design to meet that customer's specific requirement.

Administration. As in the other engineering disciplines, administration and management are logical stepping-stones for the mechanical engineer. As engineers gain more experience and show an aptitude for supervising and coordinating activities and people, they gradually find themselves with more people to supervise and more responsibilities. At this point, the day-to-day technical aspects of the job are replaced with human problems. The engineer will be guiding, formulating policy, coordinating, and interacting with people—not machines.

The preceding descriptions are not intended to be complete, but rather they give a summary of the sorts of things a mechanical engineer might be doing. While no examples were given, the field of heating and air-conditioning is also part of the mechanical engineering field.

Mechanical engineering demands an aptitude for and interest in the physical sciences and mathematics, and it requires the ability to apply these interests to benefit society and meet its needs.

Settings in Which Mechanical Engineers Work

All large industries employ mechanical engineers. Traditional industries for mechanical engineers have included the automotive, industrial machinery, utilities, chemical, computer, manufacturing, mining, and petroleum industries. However, mechanical engineers are also employed in such industries as publishing and printing, oceanographic, textile, pharmaceutical, apparel, soap and cosmetic, electronic, paper and wood products, and rubber and glass.

In addition, mechanical engineers do research and teach at colleges and universities. They also work at the federal, state, and local government level and for consulting engineering firms.

Training and Other Qualifications

Mechanical engineers follow a very traditional engineering education process at the undergraduate level. Courses include the following.

Basic science. Mathematics, physics, and life sciences. These provide a foundation for all engineering and technical courses.

Engineering sciences. Solid mechanics, fluid mechanics, thermodynamics, heat transfer, systems and controls, materials, electricity, and magnetism. In addition, some course work may be offered or required in the electrical and material engineering fields.

Design manufacturing. An introduction to the process of joining ideas, imagination, and modeling to create components and systems.

Communications. English, graphics, and computer languages.

Humanities. Courses from one or more of the following: literature, sociology, history, psychology, economics, and philosophy. These courses are designed to round out engineers and better prepare them for their role in society through knowledge and understanding of their culture, themselves, and one another.

Outlook for the Future

According to the Bureau of Labor Statistics, mechanical engineering is expected to be one of the fastest growing engineering fields of the future. The fact that other engineers depend on the expertise of the mechanical engineer assures that the field will continue to grow and expand.

Earnings

According to the 1992 Lindquist-Endicott Report, the average monthly starting salary for a person graduating with a BS degree in mechanical engineering was $2,936.00. For those graduating with a master's degree in mechanical engineerng, the average starting salary was $3,299.00. New Ph.D.'s in mechanical engineering reported an average monthly starting salary of $4,627.00.

Additional Sources of Information

The American Society of Mechanical Engineers
United Engineering Center
345 East 47th Street
New York, NY 10017
(212) 705-7375

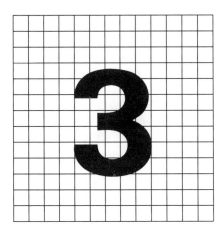

ENGINEERING SPECIALTIES

Over the years each traditional branch of engineering has developed increasingly more focused specialties. Today some of those specialties have become engineering professions in their own right.

The engineering areas in this chapter can either be pursued as a specialty area within one of the previously discussed engineering disciplines or they can be pursued as a college major at institutions that offer a more in-depth preparation for these fields.

AEROSPACE ENGINEERING

Aerospace engineering, like the entire field of aerospace, has grown far beyond its original concerns with aeronautics and space. Aerospace professionals confront many challenges and even address problems closer to earth in the areas of mass transportation, environmental pollution, and medical science. Aerospace engineers find themselves on the leading edge of technology, and their solutions to problems encountered in exploring space also provide solutions to problems closer to home.

The Nature of the Work   Aerospace engineers are involved in many different areas; however, there are about seven major divisions—each with its supporting technology—which often cross the lines of other engineering fields. These major divisions include the following.

Propulsion. The study of propulsion involves the analysis of matter as it flows through various devices such as combustion chambers, diffusers, nozzles, and turbochargers. A vehicle's propulsion system is the primary force responsible for performance.

Fluid mechanics. Fluid mechanics deals with the motion of gases and liquids as well as with the effects of the motion on bodies in the medium. Engineers working in the division of fluid dynamics called *aerodynamics* are concerned with the determination of a vehicle's shape and configuration.

Thermodynamics. This science is concerned with the relationship between heat and work. The principles of thermodynamics interest aerospace engineers studying thermal balance within vehicles, thermal effects produced by high-speed reentry into the atmosphere, and environmental control systems.

Structures. The science of structures develops advanced techniques in the areas of structural analysis, dynamic loads, aeroelasticity, and design criteria. The engineer in this field must answer two questions about any framework: (1) Is it strong enough to withstand the loads applied to it? and (2) Is it stiff enough to avoid excessive deformation and deflections?

Celestial mechanics. The science of celestial mechanics is concerned with the motion of particles in space. These particles can represent rockets, planets, missiles, or spacecraft. When engineers prepare a space mission, a major concern is determining the paths of the rockets and planets. Their calculations, facilitated by banks of computer devices, take into consideration the propulsion systems, optimum programs for fuel or propellent utilization, optimal trajectories, transfer orbits, and the potential effects of thrust misalignment.

Acoustics. Acoustics deals with the production and behavior of sound. Some of the problems aerospace engineers address include internal noise generated from stators, rotors, fans, and combustion chambers. They also study sonic booms and their effects on the urban and rural environment.

Guidance and control. Guidance and control systems automate the control, maneuverability, and path systems of a space vehicle in order to fulfill its mission objectives. Examples of systems on a more conventional level include the ILS (instrument landing system), which permits aircraft to land day or night in all kinds of weather. Similar systems also provide guidance and control for submarines.

The Settings in Which Aeronautical/Aerospace Engineers Work

Many aerospace engineers are employed in the aerospace industries which receive major contracts from the U.S. defense department. Others are employed in government agencies, particularly in the U.S. Department of Defense and the National Aeronautics and Space Administration (NASA).

Training and Other Qualifications

Experts agree that junior high school is the best time to begin planning for a career in aerospace engineering, but no later than the junior year in high school. Most colleges and universities offering programs in aerospace engineering expect the students they admit to have taken the following courses in high school:

- English
- Mathematics—algebra, geometry, trigonometry, and calculus, if possible
- Sciences—physics, chemistry, and biology
- History—including social studies

Humanities courses are also recommended.

During the first two years of college the coursework that will be taken will be very similar to what other engineering students take. However, during the junior and senior years, the program of study will be devoted primarily to design or to research and development.

Third Year
Aero-Design Program
    Applied Aerodynamics
    Elementary Structural Analysis
    Metals and Metallurgy
Aero-Research Program
    Analytical Mechanics
    Electromagnetic Fields
    Advanced Math
Common to both programs:
    Fluid Mechanics
    Heat Transfer
    Electrical Circuits
    Aeronautical Laboratory
    Nontechnical Courses
Fourth Year
Aero-Design Program
    Flight Vehicle Design
    Vehicle Stability and Control
    Structural Analysis
Aero-Research Program
    Solid Mechanics
    Flight Mechanics
    Vehicle Systems
    Trajectory Dynamics
Common to both programs:
    Gas Dynamics
    Electronics
    Modern Physics
    Aerospace Propulsion Systems
    Boundary Layer Theory

Outlook for the Future   Aerospace engineering is primarily impacted by government policies in the area of defense and space programs. While there is always a need for aerospace engineers, the availability of federal funding for research and development and production varies greatly. Cutbacks in defense spending as a result of the end of the Cold War will have a major impact on the availability of opportunities in the future.

Earnings   According to the 1992 Lindquist-Endicott Report, the average monthly starting salary for a person graduating with a BS degree in aerospace engineering was $2,766.00.

Additional Sources of Information

American Institute of Aeronautics and
Astronautics, Inc.
1290 Avenue of the Americas
New York, NY 10104

AGRICULTURAL ENGINEERING

Agricultural engineering is one of the basic engineering disciplines and probably has the closest relationship with the environment of almost any engineering discipline. To ensure tomorrow's food production, agricultural engineers work to protect today's environment.

Agricultural productivity is one key measure of an agricultural engineer's performance. As agricultural engineers develop new tools, it becomes increasingly easier and more practical to produce, process, and distribute food and fibers.

The Nature of the Work   Agricultural engineers utilize scientific principles to design systems and equipment to manage the various resources that provide food and fiber. These resources include soil, water, air energy, and engineering materials.

Agricultural engineers apply the skills of engineering to related agricultural skills. These skills are applied across the vast food production chain, from the protection of natural resources to the preservation of food products.

Natural Resources. Agricultural engineers are concerned with the proper use and conservation of our soil, water, and forests. Controlling the effects of weather, soil erosion, and water pollution are just three of the many concerns the agricultural engineer addresses on a daily basis. Agricultural engineers employ irrigation, drainage, and erosion control practices to conserve our limited natural resources. Systems to reclaim water and land are also designed and put into use by agricultural engineers.

Research, Design, and Development. Some agricultural engineers initiate and try out new designs, leading to the production of hybrids that can survive in the most hostile environments. Still other engineers work with new sources of power and energy. Producing fuels from renewable resources, such as grain, is the day-to-day activity of these engineers.

Computers are important tools of agricultural engineers involved in research and development. In providing such engineers with models of plant and animal systems, they contribute to the efficient management of the agricultural industry.

International Consultants. Agricultural engineers apply their specialized knowledge and skills to help developing nations produce adequate food supplies to meet their populations' requirements. By consulting internationally, engineers participate in the process of providing food for inhabitants of less-developed countries.

Environmental Control. Agricultural engineers provide the solution to the environmental problems of animal and poultry disease, fly breeding, and odor—all of which could endanger our food supply.

The agricultural engineer works to provide the best environmental conditions for farm animals, which in turn leads to more efficient and economical production of meat and dairy products.

Electricity is just one of the tools used by the agricultural engineer to help control the environment. Computer systems are being used increasingly to help implement engineered control systems effectively and to enhance or increase the expected yield from an acre of land or the amount of milk from a dairy cow.

Control of the soil, water, and air that nourish plants is becoming more important because of the increased production demands on a finite amount of available land. Obviously, land that has been paved over to form cities, roads, and shopping centers can no longer be utilized to cultivate and produce food. Because land is continually being removed from cultivation in this way, agricultural engineers are working with new methods of cultivation, including *hydroponics,* the emerging science that permits the cultivation and production of foodstuffs without soil; the development of specialized greenhouses; and irrigation and drainage schemes that make the most of our diminishing resources.

Agricultural Structures. Many agricultural engineers design and manufacture better structures for use on farms. Special knowledge of material-handling equipment and material flow systems is required for many complex functions performed around the farm. Structures related to these functions can include storage and processing facilities for grain, hay, feed, and silage; facilities for livestock and poultry production; and structures for processing and disposal of waste matter. Another challenge is the design of comfortable and affordable economic housing.

The Settings in Which Agricultural Engineers Work

Some agricultural engineers are employed in the increasingly technological agricultural industry. Still others run experimental farm stations and research laboratories, usually at land grant universities, that benefit the agricultural community and industry.

One of the most interesting, challenging, rewarding opportunities for agricultural engineers is serving as a consultant for a project in another country. With societies' increasing concern about world hunger, the agricultural engineer plays a key role in implementing new technologies for food production.

Training and Other Qualifications

While the field of agricultural engineering includes mechanical, civil, electrical, chemical, and other types of engineers; many are specifically trained in the field of agricultural engineering. Universities that specialize in training agricultural engineers are located at most of the land grant universities, where courses in both engineering and agriculture are offered. College programs usually require:

- calculus and higher mathematics
- chemistry and physics
- basic engineering courses, including engineering mechanics, thermodynamics, hydraulics, and energy utilization
- basic agricultural and science courses
- English
- social science
- humanities

The agricultural engineer's course requirements are not unlike those for the other disciplines, with the obvious additions of agricultural and biological science courses. In addition, the agricultural engineer will be expected to take courses in computer science and engineering design.

Additional Sources of Information

American Society of Agricultural Engineers
2950 Niles Road
St. Joseph, MI 49085-9659

AUTOMOTIVE ENGINEERING

Automotive engineering involves the design, development, testing, manufacturing, and applications of vehicles and their components. It is a field wide open to the inquisitive engineer who wishes to be involved in a broad variety of disciplines and their applications.

The Nature of the Work

Jobs in the field include the application of mechanical engineering, chemical engineering, electrical engineering, materials engineering, aerospace engineering, computer engineering, and civil engineering. Additionally, automotive engineering makes use of virtually every other field of pure or applied science and technology.

Automotive engineers not only design, develop, test, and manufacture passenger cars and trucks, they are also involved with such things as emissions safety, fuels and lubricants, construction and industrial machinery, electrical equipment, electronic systems, engine, body chassis, hydraulics, materials, occupant restraint, human factors, tires, wheels, transmissions, and suspensions.

The Settings in Which Automotive Engineers Work

Automotive engineers are employed in the automobile manufacturing industry and in automotive services. They are also employed in supporting industries such as the electronic components industry, the tire industry, the fabricated plastics and metals industries, and the transportation industry.

Training and Other Qualifications

Like all other areas of engineering, automotive engineers should begin preparation as early as junior high school by taking as much math and science as possible. Although no four-year, accredited colleges or universities award degrees in *automotive* engineering, many of the traditional areas of engineering (i.e. mechanical engineering, chemical engineering, electrical engineering, materials engineering, aerospace engineering) provide related course offerings. Colleges and universities involved in automotive engineering will usually have active student chapters of the Society of Automotive Engineers (SAE).

Outlook for the Future

The employment outlook for automotive engineers is somewhat uncertain. On the one hand, the influx of imported automobiles has made the U.S. marketplace soft. On the other hand, a revitalization and reindustrialization of the U.S. auto industry could result in more opportunities and challenges for automotive engineers as U.S. auto manufacturers seek to reestablish their dominance in the field.

Earnings

Because automotive engineers have degrees in traditional engineering disciplines, the reader should consult chapter 2 about the starting earnings of each of those areas.

Additional Sources of Information

Society of Automotive Engineers
400 Commonwealth Drive
Warrendale, PA 15086

BIOMEDICAL ENGINEERING

Biomedical engineers apply their knowledge of engineering and human anatomy to the discovery and maintenance of systems and equipment used to assist medical and other health care professionals.

The Nature of the Work   Biomedical engineers usually function as part of a medical team and are often the only professional engineer of the team. In many cases biomedical engineers provide the technical interface between the manufacturer and the user of medical equipment. Therefore, their interdisciplinary background and their engineering expertise is relied upon heavily. In general, biomedical engineers function in one of four areas: *bioengineering, medical engineering, clinical engineering,* or *bio-environmental engineering.*

Bioengineering is where engineering concepts and technology are applied to biological systems, as opposed to medical systems. Biomedical engineers working in this area design, develop, and manufacture devices and instrumentation that measure or control biological functions. Many good examples of bioengineering can be found in the NASA space program, which is concerned with monitoring the biological functioning of astronauts during space flight.

Medical engineering applies engineering concepts and technology to the development of instrumentation, materials, diagnostic and therapeutic devices, artificial organs, and other medical equipment.

Clinical engineering applies engineering concepts and technology to improved health care delivery systems in hospitals, clinics, government agencies, universities, and industry.

Bio-environmental engineering is an emerging new field. It applies engineering concepts and technology to maintain and improve the quality of the environment in order to protect human, animal, marine, and plant life from toxicants and pollution.

Although a bachelor's degree in the area of biomedical engineering is excellent preparation for medical school, biomedical engineers are not always physicians. However, their duties and a great deal of their academic preparation are similar to the duties and education of physicians. Both biomedical engineers and physicians are concerned with and study the human body. To a degree, both learn about chemical interactions in body functions. And both are concerned with the enrichment of human life. The subtle differences are in the methodology of their approaches. Biomedical engineers are in primarily supportive roles; their grasp and application of technology are utilized to supplement the treatments of a physician.

Biomedical engineers assess requirements and then design and fabricate electronic instruments to meet these requirements.

The role of the biomedical engineers is one of the most exciting and rewarding of all engineering opportunities. Premature infants, born three months too soon and weighing two or three pounds, are connected to and assisted by the electronic devices biomedical engineers are responsible for—helping the babies breathe, monitoring their vital functions, and aiding the medical team in making adjustments to a complex array of medicines and additives flowing into tiny veins—in fact, preserving the lives of such babies.

Biomedical engineers perform miracles or at least assist in their performance. Repairs to and renovations of the heart are commonplace to the biomedical engineer, who can, and often does, serve as a key member of the open-heart surgical team.

Biomedical engineers don't limit themselves to the application of electronics to human anatomy and physiology. They also integrate mechanics, fluidics, and polymer and ceramic technology to achieve a goal. For example, the artificial valve that is used as a replacement for a defective heart valve is the result of efforts by almost all the engineering disciplines. Likewise, the synthetic replacements for human body joints—for elbows, hips, and knees—are the products of the biomedical engineer integrating not just electronics but all of the engineering fields to enhance the quality of life. The following list provides an idea of the types of devices biomedical engineers design, build, operate, and maintain:

- Electrocardiographs
- Blood gas analyzers
- Artificial replacement joints such as knees, elbows, and hips
- Computer-assisted patient care devices
- Artificial hearts
- Physiological monitoring devices
- Pacemakers
- Defibrillators
- Heart-lung bypass machines
- Artificial replacement heart valves
- Computer-directed patient-care monitoring systems
- Intra-aortic balloon pumps
- Computerized Axial Tommography (CAT)
- Magnetic Resonance Imaging (MRI)
- Artificial replacement limbs

The Settings in Which Biomedical Engineers Work

In addition to the large proportion of biomedical engineers who go on to pursue medical degrees, a significant number of biomedical engineers are employed in hospitals. Not all biomedical engineers find themselves in clinical situations. For every engineer working in a hospital, there are probably five engaged in industry. These are the engineers who design and investigate new techniques and technology, taking a problem and turning it into an opportunity and then a solution.

These engineers can also be found in an academic or pure research setting, taking their in-depth knowledge of electronics and combining it with medicine to provide the means for even greater enrichment of human life.

Training and Other Qualifications

Biomedical engineers are educated not only in the traditional engineering areas, but also in the biological disciplines of anatomy, biophysics, pharmacology, physiology, neurophysiology, and organic and biological chemistry.

There are basically two types of biomedical curricula. The first leads to a bachelor's degree in biomedical engineering. The second leads to a bachelor's degree in a traditional engineering field, such as mechanical, electrical, or chemical engineering where the student elects to take courses in a biomedical option. These courses supplement the regular engineering curriculum. In addition to the required engineering course work, students take courses in biomedical engineering and biology and anatomy.

Because industrial need for biomedical engineers is concentrated at the graduate level, a large percentage of biomedical engineers continue their studies beyond the undergraduate level. A high percentage of undergraduates go on to graduate or professional school immediately after graduation. In fact, the proportion of students enrolling in medical school is generally higher than the proportion entering industry or graduate programs in biomedical engineering.

Outlook for the Future

Recent growth in the health care industry and the aging of society mean that opportunities for biomedical engineers will increase in the future. The contributions that biomedical engineers have to improving the quality of life make them an important part of the health care team.

Additional Sources of Information

American Association for Medical Systems
    and Information
4405 East-West Highway
Bethesda, MD 20814

Institute of Electrical and Electronic
    Engineers (IEEE)
345 East 47th Street
New York, NY 10017

COMPUTER ENGINEERING

Computer engineering, although traditionally a part of electrical and electronics engineering, has begun to be recognized as a separate engineering entity. Computer engineering areas include computer engineering, computer information systems, computer science, and information science.

In addition, computer engineering also crosses the boundaries of many of the other engineering disciplines, as their talents and services are necessary in developing and implementing computer systems.

The Nature of Computer Engineering Work

The opportunities within the computer engineering field cross over the dividing lines of other engineering disciplines. This is obvious in the production of a computer, which requires a collaboration by many different specialist engineers.

A computer engineer or electronics engineer designs the electronic or digital circuit. Another engineer, a printed circuit board designer, converts the engineer's design to a pattern which is then photographically transferred to a copper laminated board. The end result is a wireless chassis that will eventually become a computer.

Before this circuit can become a computer, other engineers within the computer engineering discipline must determine what it is to do and how it is to perform tasks. These engineers are software systems designers (programmers). They mate the particular hardware characteristics with applications programs or operating systems to enable the chips to function in a unified manner and produce a desired result.

Even when the systems people have finished, a computer still needs the attention of computer information science engineers to determine the manner in which the computer can best serve the intended user. Data bases are designed to store and retrieve even the most minute bits of information on a multitude of subjects. However, computer information science engineers emphasize the arrangement of input and output data rather than concern themselves with the mechanics of computing.

Systems engineers integrate the work of the hardware and software engineers and further refine and define the computing device. Their duties can require interfacing with control systems, machinery, or other external devices.

It is a reasonable assumption that a computer engineer will be working not only with electronic and electrical engineers but with virtually every other member of the engineering profession. Nothing that is done in the fabrication of a computer system can be considered the work of a single engineering discipline. Rather, it is a combined effort by all of the engineering fields, merging their specific knowledge and talents to obtain a device capable of making life easier and more productive. This effort becomes increasingly important as the United States's emphasis shifts from being a producer of goods to being a producer of information and of devices that process and manipulate information.

Training and Other Qualifications

Because computer engineering is considered a part of the electrical/electronics engineering field, computer engineers will take the same general core courses as electrical engineers. In the junior and senior years, however, the degree and amount of specialization will vary according to a student's chosen track.

The following curriculum shows the typical course titles for electrical and computer engineering and computer science and engineering:

Electrical and Computer Engineering

Math
Physics and Chemistry
Introductory Computing
Mechanics and Thermodynamics
Electromagnetic Fields
Logic Circuits and Lab
Computer Architecture and Switching
Circuits and Electronics and Labs
Energy Conversion
Linear Systems
Oral/Written Communications
Social Science/Humanities

Computer Science and Engineering

Math
Physics or Chemistry
Introductory Computing
Computer Hardware and Microcomputers
Software Engineering
Lab and Design Work
Engineering
Computer Science and Engineering Electives
Oral/Written Communications
Social Science/Humanities

Computer science and engineering programs generally have fewer courses in physics or chemistry. Instead of mechanics, thermodynamics, and energy conversion courses, the curriculum contains more electives in numerical methods, database design, operating systems, and artificial intelligence.

For More Information

American Federation of Information
   Processing Societies Inc.
1815 North Lynn Street
Arlington, VA 22209

American Society for Information Science
1010 16th Street, NW
Washington, DC 20036

Association for Computing Machinery
11 West 42nd Street
New York, NY 10036

Association for Educational Data Systems
1201 16th Street, NW
Washington, DC 20036

IEEE Computer Society
1109 Spring Street
Silver Spring, MD 20910
                 or
10622 Los Vagueros Circle
Los Alamitos, CA 90720

Instrument Society of America
Box 12277
Research Triangle Park, NC 27709

Association for Women in Computing
407 Hillmoor Drive
Silver Spring, MD 20901

ENVIRONMENTAL ENGINEERING

Environmental engineering is not as new a field as one might think. Its history can be traced back to the 1800s and to the field of sanitation engineering. Today environmental engineers play a vital role in working to reduce the pollution and toxicants in our air, water, and ground in order to preserve a better quality of life for all living things.

The Nature of the Work   Environmental engineers work in many capacities. Some work as researchers, designers, planners, and operators of pollution control facilities. Others teach in universities or work as government regulators.

The major areas in which environmental engineers work include: air pollution control, industrial hygiene, radiation protection, hazardous waste management, toxic materials control, water supply, wastewater management, storm water management, solid waste disposal, public health, and land management.

The Settings in Which Environmental Engineers Work

In addition to all types of private industries and businesses, environmental engineers are employed by private engineering consulting firms as well as federal, state, and local government agencies. They are also employed by universities and by research firms and testing laboratories.

There are international opportunities for environmental engineers, particularly in Eastern Europe. Because there is a strong relationship between pollution and population, many opportunities for environmental engineers exist where there are large concentrations of people.

Training and Other Qualifications

The American Academy of Environmental Engineers recommends that preparation for a career in environmental engineering begin with a bachelor's degree in civil, chemical, mechanical, or environmental engineering. Courses in math, science, and engineering mechanics as well as the humanities, writing, and speaking are very important. It is also recommended that a master's degree in environmental engineering is a good investment. As in any profession, more education opens more doors. Therefore, a Ph.D. is also considered to be a good investment.

Outlook for the Future

According to both the American Academy of Environmental Engineers and the National Science Foundation, there have never been enough environmental engineers. New legislation designed to protect the environment has significantly increased the demand for environmental engineers. As the environmental problems of the world grow more complex, the demand for formally trained environmental engineers is expected to continue to increase.

Earnings

The *Wall Street Journal* recently reported that starting salaries for people with bachelor's degrees can range from $35,000 per year to $41,000 per year. In addition to being one of the fastest growing areas of engineering, it is also one of the most highly paid for entry-level people.

Additional Sources of Information

American Academy of Environmental
    Engineers
130 Holiday Court, Suite 100
Annapolis, MD 21401
(301) 266-3311

NAVAL ENGINEERING

Shipbuilding is an ancient industry—some engineers would even include Noah among naval pioneers. As long as ships have been used for travel and trade, marine industry engineers have been there to make it possible.

Naval engineering, the science of marine industry engineering, encompasses a wide range of fields and engineering disciplines. The design, construction, and operation of marine systems are the basic functions of the marine industry and the jobs of marine industry engineers. Naval engineers also design ships and other marine structures, ship raw materials and finished products, and operate and service devices that perform all of these tasks.

The Nature of the Work   The field of marine industry engineering can be classified in three distinct divisions: *marine engineering, naval architecture,* and *ocean engineering.*

Marine Engineering.   Marine engineering is concerned with the design, construction, operation, and repair of energy conversion devices and systems for marine applications.

These systems can include ship propulsion systems, cargo-moving systems, refrigeration and air-conditioning systems, and control systems. Fluid and structural dynamics, heat transfer, mechanics, machine design, and electrical engineering technologies form the basis for marine engineering. Additionally, marine engineers must have a fundamental knowledge of naval architecture.

Marine engineers in the military usually work on the combat systems found on fighting ships. These rapidly changing systems provide some of the most interesting challenges in marine industry engineering.

Combat systems combine traditional armaments, such as cannon and machine guns, with the very latest in high technology, such as missile systems and automatic defensive weapons. Other armaments can include torpedoes and weapon systems on aircraft assigned to naval bases. Engineers also work on the placement of related systems, such as radar, sonar, launchers, periscopes, missile control systems, and systems for stowing, handling, and replenishment.

Naval Architecture.   Naval architecture is concerned with the design, construction, operation, and repair of marine vehicles. A naval architect might conceptualize and develop a vehicle to meet the requirements of its owner, the ocean environment, and any interface with other transportation systems.

Fluid dynamics, structural mechanics, elements of ship architecture, and ship statics and dynamics form the base for naval architecture. Also, the naval architect must have a knowledge of marine engineering.

Ocean Engineering.   Ocean engineering is concerned with the design, construction, operation, and repair of offshore structures, systems, and related support facilities for the exploration, development, and recovery of marine resources.

Ocean engineering combines oceanography, naval architecture, marine engineering, and civil engineering. Mobile and fixed exploration and production platforms, terminals, and supply systems are examples of the ocean engineer's products.

Shipbuilding allows the engineer to see a visible result of the design work that preceded the ship. The engineer is involved in the planning and scheduling and especially in the industrial engineering activities for shop procedures and strict cost containment.

Cost reduction and improved productivity of ships allow for the lowering of the cost of ship work both on a commercial and military basis. This allows U.S. firms to remain competitive with foreign shipbuilders.

A knowledge of fabrication techniques, such as welding, is required, because a shipyard not only builds vessels but effects repairs and modifications as well.

Research and development offers engineers and scientists unique challenges and opportunities for developing fundamental theories and ideas that provide practical solutions to problems. All of the most modern tools—microcomputers, computer modeling, and computer-aided design—are used by engineers to learn more about basic problems and to present viable, practical solutions that are both economically feasible and able to be translated into production capabilities.

Research and development requires all of the skills and disciplines mentioned previously as well as a specialized knowledge of the marine industry.

Settings in Which Naval Engineers Work

According to the American Society of Naval Engineers, commercial and military ship-oriented design organizations, laboratories, shipbuilding, and commercial ship operations employ most naval engineers. Some naval engineers teach and do research at colleges and universities. The U.S. Defense Department is also a major employer of naval engineers.

Training and Other Qualifications

Naval engineering requires knowledge of a broad variety of disciplines, including:

- civil engineering for weapon structures
- aeronautical engineering for weapon form
- chemical and mechanical engineering for propulsion
- electrical and electronics engineering for power supply as well as for guidance and control of launched weapons
- physics for acoustics and electro-optics

- mechanical engineering for loading, rotating, and elevating weapons launchers

There are about a dozen colleges and universities in North America where accredited programs in marine engineering, naval architecture, and ocean engineering are offered. A list of the names and addresses of these institutions can be obtained from the Society of Naval Architects and Marine Engineers, One World Trade Center, Suite 1369, New York, NY 10048.

Admission requirements differ for various schools. It is best to contact the admissions offices of schools you are considering attending to find out their entrance requirements.

Additional Sources of Information

Society of Naval Architects and Marine
   Engineers
One World Trade Center, Suite 1369
New York, NY 10048

American Society of Naval Engineers
1425 Duke Street
Alexandria, VA 22314

NUCLEAR ENGINEERING

Nuclear engineers study the basic components of all matter, neutrons, protons, and electrons. In studying them, they deal primarily with inanimate substances. Nuclear engineers work in a fascinating field with many challenges.

The Nature of the Work

The nuclear engineer participates in a variety of engineering functions that include analysis, design, synthesis, education, management, operation, and research. This investigation takes the nuclear engineer into areas such as:

- power generation and production
- medical diagnosis and treatment
- harnessing nuclear energy to search for and recover minerals
- reducing pollution caused by using fossil fuels
- new food preservation techniques
- precise measurement and dating techniques
- desalting ocean water for human consumption

• new propulsion systems for future travel and space exploration

It is obvious from this list that nuclear engineers are expending more time finding useful applications for nuclear energy than researching, developing, and producing weapon systems.

Power generation is a major application of nuclear energy today. On a global basis, our supplies of fossil fuels are dwindling, and increased consumption of these fuels leads to shortages and pollution of the environment. Nuclear power, on the other hand, is relatively clean and, with proper controls and safeguards, as safe as—perhaps safer than—other, more conventional methods of power generation.

Nuclear engineers use their skills to produce power generators capable of taking the salt from the ocean's water economically, thereby providing safe, pure drinking water in abundance for those people living in arid communities with no freshwater sources of nourishment.

Other engineers work with radiation to provide a variety of therapeutic treatments and diagnostic tests for the world's medically needy. The x-ray machine in a doctor's office or hospital is the product of a nuclear engineer working to harness the effects of radiation to serve society. Cancer cells are bombarded by carefully controlled doses of radiation, which destroy the killer cells and spare many patients the scars and trauma of surgery.

Advanced, accurate measurement and dating techniques are also provided by nuclear engineers. Thanks to them, we can measure objects precisely, examine their makeup, and even, with specialized equipment, accurately tell the age of relics from the past.

Obviously, there are many more peaceful applications for nuclear energy than military ones. The following summaries describe some of the positive uses for nuclear energy.

Environment and pollution. Nuclear power allows us to identify the components of pollution through neutron activation analysis, radiation processing of pollutants, and detection and instrumentation techniques.

Health. Radiation, and therefore nuclear engineering, aids in new and developing techniques and equipment for diagnosis and therapy.

Space exploration. Nuclear engineers study the effects of space radiation on biological systems and components of space vehicles. They also research and develop alternate propulsion systems.

Consumer and industrial power. Nuclear engineers design, develop, and select materials for structures, fuels, and coolants. They also design biological and thermal shields for controlling nuclear reactors.

Food supply. Pasteurization, sterilization, and genetic research that improves the quality and quantity of food supplies are all made possible by the nuclear engineer. Irradiation, a relatively new technique, permits long-time storage of food without refrigeration.

Transportation. Nuclear engineers develop practical transportation systems built and designed around nuclear power. Our fleet of nuclear-powered submarines and other surface warships and the USS *Savannah*, a nuclear-powered merchant ship, are examples.

Water supply. With over two-thirds of our planet under water, some method must be developed to convert seawater to safe water for drinking and irrigation. Desalting techniques require extensive amounts of electrical power, and the nuclear engineer designs and implements power generation equipment to supply these requirements.

The Settings in Which Nuclear Engineers Work

According the the U.S. Bureau of Labor Statistics, most nuclear engineers work for the federal government. However, many are employed by electric and gas utility companies as well as guided missile and space vehicle companies. Others are employed by engineering consulting services and business services, particularly those servicing the medical industry.

Training and Other Qualifications

Preparation for a career in nuclear engineering is similar to that for other engineering disciplines. Extensive math, science, and English backgrounds are required with specialized studies in atomic and nuclear physics.

Outlook for the Future

There are many challenges for both today's and tomorrow's nuclear engineers. Some of them include:

- interaction of radiation with matter
- instrumentation and controls
- reactor analysis
- controlled fusion
- energy conversion and fuel management
- fast breeder reactors
- radioisotope applications
- safety
- environment

Nuclear engineers will probably never be without challenges or opportunities.

Additional Sources of Information

American Nuclear Society
555 North Kensington Avenue
LaGrange, IL 60525

OTHER ENGINEERING SPECIALTIES

Fire Protection Engineering

Fire protection engineers fill a vital role in today's society. People, whether at home, work, or play, must be protected from the threat of fire. The fire protection engineer is responsible for researching, designing, installing, and operating a variety of hardware and systems that can economically provide a fire-safe environment.

Fire protection engineers utilize:

- scientific knowledge based on physics, chemistry, engineering science, and human relations
- mathematics as a tool to develop and model simulations of real fires
- research as a basis for new fire safety systems and hardware design and application

Like other engineers, they also rely on computers for design and to minimize routine calculations, on judgment based on experience, and on communications skills to influence and motivate others.

The fire protection engineer is in demand in a variety of fields from public service to private industry—even in colleges and universities. A career requires a firm grounding in principles of engineering as well as specialized training and laboratory experience specifically relating to fire, its effects, and the development of flame-retardant materials.

For More Information

Society of Fire Protection Engineers
60 Batterymarch Street
Boston, MA 02110

Manufacturing Engineering

Manufacturing engineers are primarily problem solvers—project organizers and researchers seeking better ways of producing the types of products in demand.

They receive an idea from a product design engineer and translate it into reality, while keeping the product cost effective. Manufacturing engineers are responsible for the development, design, analysis, planning, supervision, and construction of the methods and equipment that enable the production of industrial and consumer goods.

Previously, most firms filled the jobs of manufacturing engineers with specialists in either mechanical, electrical, or industrial engineering; however, the current trend is to fill these positions with specially trained and skilled manufacturing engineers who can integrate all of the individual disciplines and make things happen.

The employment outlook for manufacturing engineers is quite favorable. Experts agree that in order to remain competitive in the manufacturing marketplace, specially trained engineers—manufacturing engineers—are essential. Manufacturing needs can no longer be effectively served by the other, more traditional, older engineering disciplines.

For More Information   You can get more information about a career as a manufacturing engineer by writing

> Society of Manufacturing Engineers
> 20501 Ford Road
> Dearborn, MI 48128

Mining Engineering

Mining engineers find, extract, and prepare minerals for manufacturing industries to use. They design the layouts for open pit and underground mines, supervise the construction of mine shafts and tunnels, and devise methods for transporting minerals to processing facilities.

Safety plays an important role in mining engineering. There are many stories of mine disasters and cave-ins, and today's mining engineer is as concerned with the safety of the workers as they are. The mining engineer must also consider environmental protection. With public interest in conserving resources, today's mining engineer must ensure that a project will be environmentally safe, not harmful to its natural surroundings.

There are only about seven thousand mining engineers actively employed. Their workplaces range from mines to research and design facilities where new techniques are developed and tested.

Ongoing efforts to become energy self-sufficient should spur a demand for mining engineers as we learn to extract oil from other fossil fuels such as coal. However, employment opportunities are tied closely to the environment and to processes that reduce the harmful emissions that result when many of the fossil fuels are burned to produce power.

For More Information   To learn more about a career as a mining engineer, you can write to

> Society of Mining Engineers
> American Institute of Mining, Metallurgical
>   and Petroleum Engineers
> 540 Arapeen Drive
> Salt Lake City, UT 84108

Petroleum Engineering

Petroleum engineers are involved in the exploration, drilling, and production of oil and natural gas. Their goal is to achieve the maximum profitable recovery of oil and gas from a given petroleum source by determining and developing the best, most efficient production methods.

For years, when a drilling crew reached oil, a gusher of uncontrolled oil was permitted to extend hundreds of feet into the air. Today, with a diminishing source of oil, wells are no longer allowed to produce these impressive but wasteful gushers. Rather, every drop of the precious commodity finds its way into the pipelines and then to a refining facility.

Only a very small portion of the oil and gas trapped under the earth will flow under natural forces, so the petroleum engineer must develop systems and equipment to pump or force the resource from hundreds of feet below the earth's crust.

Petroleum engineers are usually employed in the petroleum industry and closely related fields. They may also work for companies that produce drilling equipment, supplies, and refining processes.

For More Information

Write to

Society of Petroleum
    Engineers of AIME
6200 North Central Expressway
Dallas, TX 75206

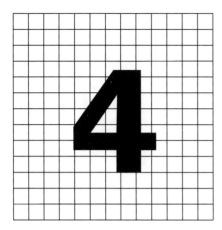

ENGINEERING TECHNOLOGY

Engineering technology constitutes a wide range of skills and methods. Engineering technologists work closely with engineers. They assist engineers in planning and implementing their designs and inventions. To understand engineering technology and what it is, let's take a closer look at the difference between the engineer and the engineering technologist.

In their jobs, technologists and engineers often perform similar tasks; however, the engineer is more often concerned with developing the overall plans and designs, while the technologist is more concerned with the implementation or completion of a specific part of a plan or design. Put another way, the engineer is concerned with the application of scientific knowledge, and the engineering technologist is concerned with the actual performance and completion of the application developed by the engineer.

The American Society for Engineering Education has defined engineering technology in the following way:

> Engineering technology is that part of the technological field which requires the application of scientific and engineering knowledge and methods combined with technical skills in support of engineering activities; it lies in the area between the craftsman and the engineer in the part closest to the engineer.

The technologist is concerned with achieving practical objectives through the application of procedures, methods, and techniques that have been proved by experience over the years. In developing plans and designs to solve complex problems, the engineer cannot develop a single, obvious best answer or plan. In solving technical problems that are sections or subsets of a large engineering plan or design, the technologist may be able to achieve a unique or specific solution.

The engineer must consider many nontechnical factors in developing plans and designs, including legal restraints, social impacts, economic factors, and aesthetic considerations. In solving technical problems, the technologist generally is not faced with such constraints and is able to concentrate on the physical and economic factors of the problem.

The engineer must exercise judgment in solving many problems to obtain the optimum benefit for society. In general, engineering technologists are not called upon to make such complex judgments, but they often must estimate and approximate conditions that cannot be completely known.

TECHNOLOGY SPECIALISTS

With the tremendous growth in technology and the increasing complexity of the applications of scientific and empirical technology, it has become increasingly necessary for technologists to specialize in one of the many distinct branches of engineering and technology. At the same time, there has arisen a separation in which certain people in a field concentrate on the development of designs and plans for the accomplishment of a given objective while another group of specialists concentrates on the practical application and implementation of those designs and plans. Hence, there is a need for both engineers and engineering technologists.

Finally, the individuals who are concerned with the practical applications—the technologists—have been divided into two classes, *technicians* and *technologists*. In general, these terms are used to designate persons with differing levels of education or experience in their field. The term *engineering technician* is applied to a person who has graduated from a two-year technology curriculum and has obtained an associate degree in applied science or engineering technology. The term *technologist* usually is reserved for the graduate of a four-year Bachelor of Engineering Technology program or for the technician who has gained wide experience in the field through many years of practice in engineering technology.

SPECIALTIES IN ENGINEERING TECHNOLOGY

The field of engineering technology covers a broad spectrum of activities, and the engineering technologist specializes in one branch of engineering or another. In this section we will take a look at the four main branches of engineering technology—civil engineering technology, mechanical engineering technology, electrical engineering technology, and chemical engineering technology.

Civil Engineering Technology

The most prominent activity in the civil engineering field is structural design. Technologists involved in this activity are concerned with some phase of the design of buildings, dams, and bridges. Structures must be designed and constructed to withstand their own weight as well as such natural forces as earthquakes and winds, and they must be suited to the environments in which they are built. The need to accommodate extremes in climatic and environmental conditions constantly presents new problems and challenges to the engineering team that is concerned with structural design and construction.

Civil engineering technologists often are employed directly in supervising and monitoring the construction of various facilities. In these activities, it is necessary to see that the structures under construction are built exactly according to the plans. Additionally, the technician and technologist often are charged with maintaining the quality of the materials used in construction. The engineering technologist working in construction also supervises the use of such heavy equipment as trucks, cranes, earth movers, concrete mixing and placing equipment, and other devices and machines.

Another key occupational area of the civil engineering technologist is the construction and operation of transportation facilities such as highways, airports, and railroads. The technologist may be involved in the initial planning phases for such facilities, assisting the engineer in predicting the growth of population, the volume of anticipated traffic, potential future problems, and possible alternate locations for transportation facilities. In these studies, the technologist will be called upon to give full attention to the environmental impact of construction and operation of transportation facilities.

As in other fields where the construction of facilities is a necessary and important activity, in transportation engineering the technologist is involved in constructing facilities efficiently under varying conditions of terrain and climate. He or she will be involved in surveying and mapping and in the supervision of construction. Finally, he or she also may be involved in the analysis of transportation systems to ensure the maximum efficiency of completed networks.

One of the oldest and most important areas of interest for civil engineers and technologists is that of hydraulics—the management of water resources. Included in this area are the collection, control, use, and conservation of water. Projects for flood control, drainage, reclamation, and irrigation are planned and designed by civil engineers and technologists who are specialists in hydraulics, as are navigation projects, water storage projects, and hydroelectric power plants.

One of the most important areas of activity for civil engineering technologists specializing in hydraulics is their work in assuring safe drinking water and effective sewage and wastewater disposal systems. With populations becoming more concentrated and modern industry growing rapidly, the amounts and kinds of pollutants poured into rivers and streams and other areas of the environment assure that engineers and technologists working in the area of water supply and wastewater treatment will find new challenges of ever greater complexity.

In recent years, there has been a population shift from the country to the city. Civil engineering technologists working in construction, in structural design, in transportation, and in water management are concentrating many of their activities in population centers, solving the problems caused by this redistribution of people. Some civil engineering technologists devote their time primarily to working with city planners and urban development agencies, assisting them in formulating plans for growth and the management of urban areas.

Mechanical Engineering Technology

As the name indicates, the mechanical engineering technologist quite often is concerned with the development of machines. All of the power transmission elements of machinery must be joined together to form the most efficient production systems at the lowest possible cost. In developing the new machines that our complex society demands, the designer also must become involved with the development of machine tools—those tools that are used to rapidly manufacture great numbers of parts for other machines. Machine tools include mills, grinders, shapers, and lathes. Quite often the technologist is involved in designing and manufacturing these tools as well as in supervising the use of them.

In order to accomplish the goals of transporting people and goods and of manufacturing objects, mechanical engineering technologists assist the engineer in harnessing various sources of energy. In this, they are concerned with engines and other devices for producing work from all possible sources of energy in the most efficient manner possible. The engines they are concerned with may transform the energy of falling water into mechanical or electrical energy, or the engines may run nuclear power plants for the conversion of energy released by the splitting of atoms. Engines also may be utilized to convert chemical energy into fuel, as in the common internal combustion engine. In all these activities, mechanical engineering technologists are concerned with obtaining reliable and efficient performance from engines and in monitoring their performance. The prevention of air and water pollution by such engines is also of concern to mechanical engineering technologists.

One of the principal uses of engines is in the safe and economic transportation of people and goods. Mechanical engineers and technologists are deeply involved in the development of new power plants and mechanical configurations for automobiles and trucks. Because of the increasing scarcity of cheap petroleum fuels, great attention is being given to development of more economical engines and power plants for vehicles. Mechanical engineering technologists are employed to assist in the design and perfection of automotive vehicles and in the supervision of vehicle construction. Additionally, many technologists in this field are employed in the search for means to reduce air and noise pollution associated with automobiles and trucks.

Some mechanical engineering technologists are involved in developing other means of transport, including rapid transit systems and space exploration systems. Technologists can find employment with the manufacturers of aircraft and space vehicles, as well as with industries that produce such forms of transportation as elevators, conveyors, monorails, and escalators.

Mechanical engineering technologists are also concerned with the development and operation of heating and ventilating systems, including solar energy systems. Fuels are converted in various ways to obtain the energy contained in them—and this energy must be supplied to living spaces.

Electrical Engineering Technology

A student who is interested in a career in electrical engineering technology may wish to consider working in the field of electrical power generation and transmission or on the use of electrical power in the development of communications systems or new and more efficient lighting systems.

The development of communications technology as a distinct area within electrical engineering technology has been very rapid during the last fifty years and continues at a great pace. Telecommunications is wide ranging; it makes possible a local phone call or a long-distance call, as well as the relay of television signals from a satellite to the earth. The sequence of communications devices based upon electricity and electronics includes the telegraph, the telephone, the wireless set, the radio, motion pictures with sound, telephoto transmitters, and television. These devices have been developed and manufactured in greater complexity with more reliability and at lower costs with each succeeding year. Electrical engineering technologists can participate in the development and testing of new communication devices as well as in supervision of the manufacture of already developed devices. Additionally, technologists specializing in communications may be involved in the generation, transmission, and reception of electromagnetic waves through radio and television stations and other communication facilities.

In addition to the direct use of electricity to power industry and to furnish light and heat in homes, electrical phenomena are being utilized in ever greater sophistication and complexity to control and monitor industrial processes, living conditions, and many other processes. Many of these sophisticated instrumentation systems are based upon electronics.

Electrical engineering technologists who are involved in electronics assist in the design, manufacture, and use of various types of vacuum tubes and solid-state electronic devices. In addition, they can assist engineers in designing and testing new electronic devices intended for communications or instrumentation.

Electrical engineers have become instrumental in the design and maintenance of electronic computers. They design computer circuits, plan computer layouts, and formulate mathematical models of technical problems that can be solved by computers. The electrical engineering technologist is almost certain to become involved in some way in the design, manufacture, or use of such devices if this area is of interest to him or her. The use of microprocessors and minicomputers in this field is one of the most rapidly growing areas of technology and promises a growing future, too.

Chemical Engineering Technology

Chemical engineering technologists are involved with the preparation, separation, and analysis of chemical substances. They often study the composition and changes in composition of natural and synthetic substances. In these activities, chemical engineering technologists rely heavily on a background in chemistry. However, their activities are not limited to the preparation and analysis of chemical substances. The chemical engineer, as opposed to the chemist, is concerned with the maximum utilization of raw materials when mass-producing substances via technology that controls chemical and physical processes. Technologists work with chemical engineers in the development of new products, the design of new processes, and the planning and operation of chemical plants. They may assist chemical engineers in the manufacture and analysis of such chemicals as salts, acids, or alkalis, all of which are used in great quantities in modern manufacturing processes.

Technologists in chemical engineering also may be involved in the refining of such natural materials as petroleum and rubber. Petroleum is utilized as a fuel in such forms as natural gas, gasoline, kerosene, and fuel oil. The chemical engineering technologist may be involved in the refining and purifying of petroleum fuels, as well as in the manufacture of chemicals from petroleum (petrochemicals). This is a growing field in which chemical engineering technologists are being employed in ever greater numbers.

Another growing field in the chemical engineering industry is the synthesis of biochemicals, produced in nature by plants and animals. The chemical engineering technologist working in this area is interested in developing such chemicals in great quantity, at a reasonable cost, and with a high degree of purity. In other words, the technologist is employed in trying to reproduce, in full-scale manufacturing plants, the biochemical processes which occur in nature.

In all of the activities mentioned, the technologist is involved in the production of a given chemical substance through control of a chemical and physical process. Because of the importance of process control, many technologists also are employed in the study and perfection of basic chemical and physical processes. Within engineering plants, they assist in the control and perfection of chemical reactions. They are also concerned with the design of separation equipment and the development of control systems for the separation process.

One of the most important applications of separation combines the work of the chemical engineering technologist with that of the civil engineering technologist. These professionals collaborate in the use of separation operations to purify drinking water and to treat sewage waste. In utilizing various such operations, chemical engineering technologists try to make reactions proceed as rapidly as possible with the lowest input of energy, to achieve the greatest efficiency and the lowest cost.

EDUCATION

Within the last twenty years, a new type of engineering technology program leading to a Bachelor of Technology degree has been established in institutes, colleges, and universities throughout the United States. These programs developed for a number of reasons, the most important of which appears to be the increasing complexity of modern technology. The applications of science in today's world have become so varied and complex that it is now necessary to acquire a high degree of specialization. Thus, many two-year engineering technology programs have been expanded into four-year programs which compare in technical content to the four-year engineering programs that existed in this country a generation ago.

Students can elect to earn a bachelor's degree in engineering or one in engineering technology. The baccalaureate engineering graduate very likely will hold a position in research, conceptual design, or systems engineering. The holder of a degree in engineering technology probably will be working in operations, product design, product development, or technical sales. The associate engineering technology graduate very likely will hold a position in support of an engineer's work.

A program leading to an engineering degree consists of courses in physical science, engineering science, and advanced mathematics through differential equations. The course of study leading to a bachelor's degree in engineering technology includes courses in technology, applied science, and mathematics through differential and integral calculus. The associate engineering technology program offers courses in science, skills, and mathematics through algebra.

We have seen, then, that at the present time two options in engineering technology education are open to high school graduates: the two-year associate degree program, to become an engineering technician; and the four-year Bachelor of Engineering Technology program, to become an engineering technologist. These degrees are offered in several types of institutions, including technical institutes, community and junior colleges, and universities. Approximately 150 colleges offer programs leading to associate degrees in engineering technology, and nearly 100 colleges and universities offer programs leading to the Bachelor of Engineering Technology degree. A list of accredited programs in all engineering fields is available from the American Board of Engineering and Technology, Inc., 345 East 47th Street, New York, NY 10017.

For More Information   As a college student, you can join a number of social, honorary, and professional organizations affiliated with engineering technology. Once you are a practicing technologist, you can become certified in recognition of your capabilities. Practicing engineering technologists can join professional organizations such as the American Society of Certified Engineering Technicians (ASCET), 2029 K Street NW, Washington, DC 20006.

IS ENGINEERING FOR YOU?

Now that you have read about the variety of jobs and the challenges that are available through a career in engineering, you need to determine whether or not this is the right career field for you. To do this, it will be necessary for you to examine your *personality, interests,* and *abilities.*

WHAT TYPE OF PERSONALITY DO YOU HAVE?

A person interested in an engineering career should consider whether he or she is able to relate well to others. It is essential for an engineer, who works as a member of a large team, to get along well with other people and to be able to communicate effectively with them. In considering a career in engineering, a person should examine his or her own past performance in working with others in groups to accomplish specific goals and objectives.

Think broadly here. Teamwork does not only mean class projects, it means team sports. Teamwork means student government. Teamwork means school, community, church groups, and clubs.

Do you participate in these types of activities? Do you enjoy them? Do you contribute your ideas and carry your fair share of the work load? Do you assume a leadership role?

Self-discipline is another important characteristic. Can you stick to a task and get it accomplished? This does not only apply to jobs that you may hold. It also applies to your academic classes.

Are you willing to take more challenging courses and give them your full concentration? Do you work to achieve the kind of grades that you are really capable of? Do you ask for assistance when you do not understand concepts presented in class?

These are the types of personal characteristics that many successful engineers exhibit in their professional life. You may already show signs of them in the way you approach school and work now.

WHAT ARE YOUR INTERESTS?

To discover where your interests lie, it may help to review your life to see what ideas or activities have aroused your curiosity. With respect to engineering, it is important to consider any interest you may have had in technical and scientific developments.

Are your hobbies related to scientific and technical activities? Do you like mathematics and mathematical puzzles? In your previous reading about various career fields, have you found the descriptions of engineering to be interesting or boring? If you have visited a manufacturing plant, a laboratory, or a design office, were you interested in the activities of the technical personnel there? Have you ever posed questions about their work to engineers in any of the fields discussed earlier? Have you tried to learn as much as you could about the area of engineering that interests you the most? The answers to these questions can go a long way toward indicating whether you truly have an interest in engineering.

WHAT ARE YOUR ABILITIES?

In addition to a serious interest in the field of engineering, a prospective engineer must have certain capabilities in order to be successful. Among these capabilities is a mastery of mathematics and the application of mathematical principles in the solution of practical problems.

You should ask yourself: "How good am I at stating problems in quantitative terms and in obtaining quantitative solutions? Can I translate real situations into abstract terms or symbols without becoming confused? Do I understand the various number systems used in mathematics?"

In addition to asking these questions about mathematics, you should examine your interests and successes in such related fields as physics and chemistry. Do the principles of physics interest you? Can you find any application of these principles in the real world? Can you understand concepts such as force, mass, acceleration, leverage, gravitation, and other abstract physical principles? Are you interested in studying the chemical formulation of compounds and the way in which various elements combine in nature? It should be easy to answer these questions by reviewing your own academic record in high school. High scores and excellent grades in mathematics, physics, and chemistry usually indicate that a person has the capability to undertake engineering studies with a significant chance of success.

Other capabilities are also extremely helpful, if not necessary, for a successful career in engineering. One of these abilities is the power to visiualize in concrete terms what is described in words. A person interested in a career as an engineer should consider whether he or she is able to describe in words, and to illustrate in sketches, complex actions or processes. For example, would you be able to describe in words and illustrate with a few simple sketches the mixing of fuel and air in the carburetor of an internal combustion engine? Could you show in a simple sketch the way an airplane wing works to lift the aircraft? The ability to communicate in words, both orally and on paper, is very important for success as an engineer.

WHAT SHOULD YOU DO IN HIGH SCHOOL?

During your high school career it is very important that you take as many mathematics classes as possible. These courses should be algebra I & II, geometry, trigonometry, and calculus. In addition, you also need to take as many science courses as possible. This includes biology, chemistry, and physics. Chemistry and physics are required courses in undergraduate engineering programs; therefore, you should take as many chemistry and physics courses as possible in high school, particularly honors and advanced placement courses. Four units of English are also required. Social studies and foreign language courses are necessary for admission to collegiate engineering programs. Computer and economics courses are also recommended. Your extracurricular activities during high school are also important. Math and science clubs will demonstrate strong and consistent interestes related to engineering. But it is important to participate in other activities such as athletics, service organizations, and cultural activities.

You can obtain more information about careers in engineering from counselors and teachers. You can also obtain information from the local chapter of the National Society of Professional Engineers. The Junior Engineering Technical Society (JETS) exists for the purpose of furnishing information to high school students interested in careers in engineering, technology, science, and mathematics. One of its primary activities is the sponsorship of JETS chapters in junior highs and high schools. These chapters, in turn, sponsor extracurricular clubs, under the supervision of a faculty advisor assisted by a volunteer professional engineer from the community. The typical JETS chapter or club holds regular meetings at which members explore various aspects of the field of engineering, technology, science, and mathematics. The club may visit local industries, consulting firms, and government agencies to discuss their interests with practicing engineers and scientists. Programs, projects, and other group activities are part of most clubs' programs.

Information concerning JETS can be obtained from any local chapter or from the JETS National Office, 345 East Forty-seventh Street, New York, NY 10017.

HOW TO CHOOSE A COLLEGE

College catalogs are important tools in comparing and evaluating engineering education programs. It is important to learn how to read them so that you will be able to ask good questions and get full information on the requirements and expectations of each program.

The first step is to look at entrance requirements. Many colleges and universities have minimum entrace requirements. If you are not sure if you meet the entrance requirements, you should speak to the admissions officers of the colleges or universities in which you are interested. Admissions officers will be willing to discuss how they evaluate applications.

The second step is to look at the engineering majors and specialties that are offered at each college or university. Do the colleges and universities that you are considering offer the type of engineering major in which you are interested? Will you have to maintain a certain grade point average in order to declare a major in your area of interest? If a certain grade point average is required to declare a major, what percentage of entering students are able to maintain the average required for the major in which you are interested? If you are unsure about your major, does the school provide a wide enough range of options from which to choose?

The college catalog should tell you the cost of tuition, fees, room, and board. However, it is important to keep in mind that these will not be your only costs. You will buy books, travel to and from school, buy personal supplies, pay for entertainment expenses, and incur other day-to-day costs. Therefore, it will be important to look at the availability of financial aid. The catalog will give you information on grants, scholarships, loans, ROTC, and work-study programs that are available to help you finance your education, if you are eligible. However, it is always advisable to meet with the financial aid director to discuss your specific needs.

"College Nights" are an excellent opportunity to gather information on admissions requirements, costs of attendance, and financial aid opportunities. Start going to these events throughout your high school experience, and get to know the admissions people. However, nothing substitutes for a visit to the college campus. It is important to see and feel the campus environment as well as the engineering environment. That is the best way to determine if there will be a good fit between you and the college you choose to attend.

HOW TO MANAGE YOUR ENGINEERING CAREER WHILE IN COLLEGE

Your career in engineering will not start when you leave college. It will actually start while you are an undergraduate student. When you pick a major (mechanical engineering, chemical engineering, biomedical engineering, etc.) you have taken the first step in your career. It is not an irreversible step but it is a step!

Every engineering program will require that you take certain courses both in your major and outside of your major. In addition, every program will require you to take elective courses. These are courses that you get to select. In some programs you will pick elective courses from an approved list, in others you will choose courses that follow a theme, and in still others you will be able to choose any courses that you wish. These opportunities to select courses give you an opportunity to continue to shape your career.

For example, you may use your free electives to gain more computer expertise, or a better understanding of how government policy can impact an engineer's work. You may decide to use your free electives to develop your foreign language skills so that you can practice your engineering skills in a global economy. Careful selection of your elective courses can make you more attractive to future employers and/or graduate schools.

The engineering experience that you gain during your college career should extend beyond your classes and labs. Internships and summer jobs in an engineering setting can be important to your career. However, many engineering employers place more value on cooperative education experience.

Cooperative education (co-op) is an engineering educational approach that began at the University of Cincinnati in 1906 when Professor Herman Schneider's research indicated that it was too costly for engineering employers to train new engineers after graduation. For nearly 90 years, thousands of engineers have alternated periods of paid work experience related to their major with periods of academic course work. These cooperative engineering education graduates completed their undergraduate education with a four-year degree and more than one year of increasingly responsible engineering experience. Many collegiate engineering schools offer coopertive education programs. Generally these are five-year programs, and numerous studies have indicated that the fifth year is well worth the investment. Year after year surveys have shown that the starting salaries of co-op engineering graduates have been significantly higher than those of non co-op graduates. However, most co-op graduates will say that the real benefit of the program is the experience they gained and the increased awareness of what they enjoy and want to do as professional engineers.

The co-op programs at the University of Cincinnati, the University of Detroit, Drexel University, Northwestern University, Purdue University, Georgia Institute of Technology, Northeastern University, and Virginia Polytechnic Institute and State University (Virginia Tech) are among the oldest engineering co-op programs in the country.

BEYOND THE UNDERGRADUATE DEGREE—WORK OR GRADUATE SCHOOL

When the bachelor's degree in engineering is awarded, the recipients have many options open to them. If you receive a BS degree in engineering you can choose to pursue your engineering career in a wide variety of areas such as: industry, business, consulting, marketing, management, government, research, teaching, sales, and the military.

In addition to the numerous employment options that engineers have, many engineers eventually pursue study beyond the bachelor's degree. Some go on to medical school or law school. Others obtain graduate degrees in business or management. However, many pursue master's and doctoral degrees in engineering disciplines.

GETTING STARTED IN YOUR CAREER

The first few years after completion of an engineering degree should be considered a period of apprenticeship to the profession. This does not imply that advancement may not be expected during these years. Actually, there is much that new engineers can do during this period to qualify for future professional recognition. It lies within the power of every individual to improve his or her chances for advancement and to be prepared when the opportunity arises to better himself or herself. This preparation can begin in high school and should be a continuing process.

Let us assume that you have carefully selected and graduated from a good school and that your technical preparation is the best that the school's facilities and your capabilities permit. Additionally, let's assume that you possess all the personal qualifications we have been discussing throughout this book as essential for engineers. What, then, are the factors that determine how quickly you will advance in your profession? There is, of course, a certain degree of chance that enters into the timing of certain openings; however, there are several factors to consider that could influence how well you can make the most of each opportunity.

Human Relations Engineering demands a clearer understanding of human relations and a greater degree of adaptability than many other professions. As an engineer, you will be faced with the need to reconcile a highly technical science to human reactions and limitations. The engineering project that is purely a matter of cold scientific or mathematical fact and that does not require consideration of human relations is rare indeed. Even when the facts are clear and incontrovertible, there are usually basic assumptions to be defended and alternative methods to be accepted or discarded. Except for minor projects, cooperative effort is essential in the conception and execution of every engineering undertaking.

The ability to work with others really begins at home and in early school experiences and develops all through college. It is applied in one's career, beginning almost immediately. Cultivation of this ability in college comes in the form of classwork and laboratory projects, where contacts with other students and faculty are important, and in extracurricular activities.

Cooperating and taking the lead. At the professional level, the engineer who is to be a success will already have learned to cooperate with others and, when occasion offers, to take the lead. Many of the laboratory and field projects in an engineering school require teamwork in planning, execution, and reporting. These projects are, of course, under faculty direction which corresponds closely to the supervision young engineers receive in their work. The difference, however, is that in school, all members of a team have had about the same training and are not too far apart in ability; the situation in actual engineering is much different.

Meeting competition. In practice, you will quickly meet competition, often in the form of older engineers with more experience. This can be all to the good and usually can be turned to advantage. The give and take of technical information and ideas, acknowledgment of the contributions of various members of the organization, and recognition for individual accomplishments all combine to mold character and to determine your status as an engineer.

Out of this competition will come future executives; leaders in specialized fields; and the good, solid production workers in the profession. For example, the president of a large corporation decided as a young engineer to curry favor from above and to dominate each situation where opportunity occurred. This was a carefully worked out plan, and because it was combined with good technical ability and good judgment, the rise of the individual was meteoric.

Consideration of differences. On the other hand, there will also be the plodders. Some of these fall into their particular class through choice, perhaps because of technical interest in details, a desire for perfection, or an interest in seeing things through to their absolute conclusions. One engineer, a careful student, was so free in contributing time and detailed information to others that one after another of his associates passed him in the organization, often largely as a result of his aid. Rather than being frustrated by this situation, the engineer seemed to gain complete personal and professional satisfaction from helping others to advance.

The ability to work with others is important at all levels and contributes much to making work situations more pleasant. Treatment of subordinates, associates, and supervisors can and should be on a friendly, frank basis of courtesy and equality. Occasionally, however, instances of jealousy and overbearance, of climbing over the backs of others, may be encountered. Such situations create difficult problems in human relations and may give rise to injustices. You should be aware of the existence of such problems and should be prepared to deal with them in a professional manner.

Aggressiveness

A degree of personal aggressiveness is needed to overcome minor career obstacles. You must, however, temper this aggressiveness with restraint. Good judgment about where and how to push a project is just as important as the push itself. As many engineering students suffer from ill-advised aggressive action as languish because of minor discouragements.

Every engineer has faced discouragements which threatened to bog down his or her endeavors. Without a spark of an aggressive spirit, these low points can slow you down and may even stop your production. If you don't combat discouragements and frustrations with some degree of aggressiveness, your professional performance, and consequently your career advancement, may be less than satisfactory.

Personal Integrity

The quality of personal integrity is much more than just a matter of honesty. As an engineer, your application of the laws of nature in the light of your experience calls for combined judgment and right thinking. Every technical decision must reflect your integrity and must be a stepping-stone to a sound professional reputation.

Just as your associates judge your professional integrity, the public judges the integrity of your profession. You must strive for accuracy in your professional opinions and soundness in your nonprofessional judgments. In dealing with a client or employer, professional standing rests very much on personal integrity. Confidences, which frequently are necessary in the development of engineering projects, may involve questions of finance or policy. Obviously, these require the highest professional integrity. It is a tribute to the engineering profession that so few instances arise in which the integrity of the individual can be challenged.

Of a less serious nature than unethical conduct, in the eyes of some, is distortion of a case by unduly emphasizing some pertinent points and suppressing others. For instance, enthusiasm for one phase of a project may lead to unintentional distortion if certain factors are championed at the expense of other, equally important details. The same standard of integrity should apply here; suppression of the truth is always dangerous, particularly in as exacting a field as engineering. The degree of emphasis to be placed on certain factors may become a matter of judgment.

Judgment

In engineering, as in most other human activities, personal judgment plays an important role. Judgment is important not only for personal advancement but also for determining the value of, or even the success of, each individual project.

Technical judgment. In a typical engineering undertaking, study and analysis of the technical features of the project quickly present alternatives which require the engineer to make an outright choice of method or to decide upon the degree of emphasis to be placed on one phase or another. Engineering judgment must be exercised in planning an investigation and in continuously charting the path that is to be followed.

Since engineering is the practical application of the laws of nature, the value or anticipated value of a project is an important factor in planning development. For this reason, sound judgment as to which projects should be carried forward and which ones should be dropped is extremely important.

A sense of balance. A prominent artist once said, "A true artist is one who knows when to stop." Similarly, a good engineer must know when to stop a given line of investigation and get back to the main theme. Many an engineer is so fascinated by the interesting results of research that he or she fails to weigh fairly their value in terms of the complete project.

Responsibility

Insofar as state laws can protect the public, every possible safeguard is taken to limit the right to practice engineering to those who have established their sense of responsibility. Licenses to practice are issued only after all technical requirements have been met and the applicant has demonstrated his or her qualifications during a probationary period. He or she must be vouched for by competent engineers who are acquainted with his or her professional work. These sponsors cannot conscientiously recommend the candidate for license unless they are satisfied about his or her reliability.

Safety. A spectacular example of the vital role that safety plays in engineering is the case of "Galloping Gertie," the Tacoma Narrows Bridge. This experimental departure from accepted practice in suspension bridge design collapsed November 7, 1940. Fortunately, no one was hurt. At the time, its designer was trying to prove that the vibrations to which it was subject could not be dangerous.

This example demonstrates how important safety considerations should be to the engineer. It should be mentioned here that the ECPD's "Code of Ethics of Engineers" states, "Engineers shall perform services only in the areas of their competence."

Thoroughness. Another factor that can influence your rate of advancement is your ability to assume responsibility in small matters. It is usually annoying for a supervisor to be called in to make petty decisions which you, as a young engineer, should be making for yourself. Making decisions as questions arise tends to expedite the work and at the same time helps to set a pattern that should lead to greater responsibilities.

Leadership   It is the general belief that advancement is almost synonymous with leadership. In some organizations, this undoubtedly is true; in the case of the higher-paid executive positions, the linkage is very close. These positions frequently mean, however, that the engineer has grown out of his or her profession, not through leadership in engineering as such but because he or she has displayed pure executive ability. In such cases, engineering has served as a stepping-stone to other opportunities. By recognizing early the advantages of leadership ability, you can do much to enhance your professional advancement. Again, though, some degree of restraint is called for in exercising leadership. You should be careful to avoid creating the feeling that you are imposing your will on others.

Developing leadership qualities.   The development of leadership qualities can begin in high school—in the classroom, in extracurricular school activities, and in social contacts outside school. In college, the opportunities are even greater, since many engineering laboratory assignments require cooperative teamwork in which leadership is essential.

Job Satisfaction   Job satisfaction is essential to personal career advancement. In engineering, personal interest in the subject, importance of the project at hand, and continuing professional recognition and participation in engineering society activities are important for job satisfaction. There are other factors that can affect job satisfaction and influence advancement.

Nature of work.   Feelings of job satisfaction or dissatisfaction can be generated by the type of work you are doing. Unfortunately, it is not always possible at the time of hire to foresee exactly what a job will entail, and many young engineers have found that all their preconceived ideas regarding the kind of work they most wanted were subject to considerable change during the first few years after school. Many have found the specialized engineering work they first selected not to their liking and have entered another engineering discipline that seemed more enjoyable and fulfilling.

Relationships with supervisors.   The morale of any organization is influenced greatly by its supervisors. As a new engineer, you will find that your relationship with your supervisor will greatly affect your career advancement and job satisfaction. Good communications and mutual confidence between you and your supervisor will greatly enhance your working situation and your long-term professional development. Good supervisors are friendly, have a sense of fair play, and are quick to recognize employees' technical abilities and accomplishments.

Working conditions.   Although some of the great advances in engineering have had their origins in crude surroundings, certainly many of the more recent engineering developments would never have been made without modern laboratory and shop facilities. Your professional advancement is not necessarily determined by the kind of plant you work in, but pleasant surroundings can contribute to your technical output and help generate feelings of job satisfaction. Hence, the close tie between plant facilities and professional recognition. Good working conditions also can help foster company pride.

Modern plant equipment is only part of favorable work conditions; working hours and the facilities available to the engineer are equally important. As an engineer, you will be constantly concerned with problems of efficiency and economy. Concern for efficiency and true economy should be reflected in the office facilities available to you. You can hardly be expected to give your best efforts on engineering projects if your own efficiency is hampered by outdated equipment and an uncomfortable work environment.

Salaries. During your first few years of engineering practice, you should take stock frequently of your salary prospects. If the entire group with which you are associated is progressing and there are indications that this progress will continue, you should consider your rate of advancement as comparable to your associates'. This may be a little difficult sometimes, as salary information usually is confidential. However, without bringing in any personalities, it should be possible for you to learn the average for the group or the salary experience of those who are a few years ahead of you. Your supervisor should be willing to discuss these salary matters in general terms or as group averages.

Recognition. If you are satisfied with your job and are performing well, you have every right to anticipate continuing recognition in the engineering profession. This may take the form of recognition for individual accomplishments in connection with your job, for contributions to the technical literature or to nontechnical subjects of interest to your profession, or for work in technical societies.

Engineering and engineering technology accreditation are available from:

> Accreditation Board for Engineering
>   and Technology, Inc.
> 345 East 47th Street
> New York, NY 10017
> (212) 705-7685

A list of colleges and universities offering cooperative engineering education programs is available from:

> National Commission for Cooperative
>   Education
> P.O. Box CAR
> 360 Huntington Avenue
> Boston, MA 02115

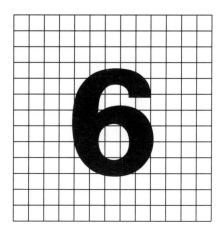

PROFESSIONAL LICENSING

Engineering is a field in which you can become a registered or licensed engineer in the state in which you work. Engineers who become registered or licensed are known as "professional engineers." They are able to put the designation "P.E." after their names. This designation conveys a level of commitment to the engineering profession that is highly valued in a number of industries and engineering disciplines.

WHAT LICENSING CAN MEAN FOR YOUR CAREER

According to the National Council of Examiners for Engineering (NCEE), there is a growing demand for engineers who have become licensed engineers. In part, this demand is due to the perception that P.E.'s are committed to the highest ethical and work standards and to their own futures and the future of the profession.

Many employers want P.E.s on their team because the designation of P.E. provides credibility to clients and to customers. In very competitive fields within engineering this can be an important advantage. For civil engineers and for engineers with consulting engineering firms the designation of P.E. is not just an advantage, it is often a requirement. In some industries the P.E. designation is highly recommended for management positions. Consequently, becoming a licensed engineer could be one of the most important career decisions you will make.

The NCEE identifies the following benefits of becoming a P.E.:

- *Promotability.* Many employers require licensure for advancement to senior engineering positions. This is particularly true when companies are engaged in internal and external partnership agreements.

- *High Salaries.* P.E.'s often earn higher salaries than nonlicensed engineers.
- *P.E. Title.* P.E.'s can sign and seal documents, and legally represent themselves as "professional engineers" to the public.
- *The Future.* Only P.E.'s are eligible to work legally as engineering consultants.

LICENSING REQUIREMENTS

It takes several years to become a P.E. The length of time actually varies from state to state as well as individual to individual. However, all registered or licensed engineers have gone through three phases to achieve this designation.

Stage 1. Every P.E. is a graduate of an ABET accredited engineering program.

Stage 2. Every P.E. has passed the Fundamentals of Engineering (FE) exam. This was formerly called the Engineering-In-Training (EIT) exam. If you see references to the EIT exam, it is the same thing as the FE exam. This is an engineering and science fundamentals test.

Stage 3. After several years of engineering experience, every P.E. has passed the Principles of Practice of Engineering (PE) exam. This is a test of knowledge in a specific branch of engineering (i.e. civil engineering, electrical engineering, etc). The PE exam tests your ability to apply engineering principles and judgment to professional problems.

TIPS FOR MEETING THE LICENSING REQUIREMENTS

If you think that you will eventually want, or need, to become a licensed engineer, you must make sure that the college or university that you attend has an ABET accredited engineering program. You need to be careful when selecting your program of study. Some departments or programs within a college or university can be ABET accredited, while others are not. Lists of ABET accredited engineering schools and specific engineering programs are available by writing to the Accreditation Board for Engineering and Technology.

The next step in the licensing process will be passing the Fundamentals of Engineering (FE) exam, formerly the EIT exam. It is administered every fall and spring by state engineering registration boards.

The Chair of the Engineering Deans Council of the American Society for Engineering Education recently stated that she believed that engineering students benefitted in the long run if they took the FE exam before graduation. This is sound advice when you consider how much knowledge you accumulate as an undergraduate. Two or three years after graduation, it can be extremely difficult to recall concepts and facts necessary to pass the FE exam.

All state boards of registration administer the same FE examination. It is produced by the NCEE. However, the dates that the exam are administered can vary slightly from state to state. It is necessary to apply to take the exam well in advance.

Engineering students are advised to take the exam during their senior year in college. Therefore, it is recommended that during the last semester or quarter of your junior year in college you contact the engineering registration board in the state or states where you plan to become licensed. They will be able to tell you when and where the FE exam will be administered during your senior year and what the application deadlines will be.

The FE exam consists of two four-hour periods. The morning exam tests comprehension and knowledge as well as evaluation, analysis, and application. The afternoon exam is composed of problem sets from seven subject areas: statics, dynamics, mechanics of materials, fluid mechanics, electrical theory, and economic analysis.

The FE exam is an open book test. However, states do vary in the amount of material that you are allowed to bring into the exam. This is important information to ask about when you contact the state board of registration regarding test dates and deadlines.

The third step in the process of becoming a licensed engineer is fulfilling your state's requirements for years of professional experience. The minimum number of years of professional engineering experience is two. However, this requirement varies state to state.

It is important to know that in many cases your participation in a college co-op program may count toward this experience. You will need to have been registered in a formal ABET accredited co-op program. The semesters or quarters that you co-op (work) will then be documented on your academic transcript for review by the registration board.

Finally, you will need to pass the PE exam. This exam will test your in-depth knowledge of a specific field of engineering. If you are a mechanical engineer, the exam that you will take will test your ability to apply mechanical engineering principles to real life problems. Likewise, if you are a chemical engineer, you will be tested on the engineering principles in that discipline.

The specific requirements of each state board of registration can be obtained by contacting them directly. For a list of the state boards and their addresses, contact NCEE.

ADDITIONAL INFORMATION

> National Council of Examiners
>   for Engineering
> P.O. Box 1686
> Clemson, SC 29633-1686

Accreditation Board for Engineering
and Technology
345 East 47th Street
New York, NY 10017-2397.

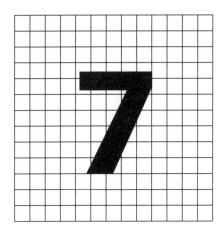

WOMEN IN ENGINEERING

Engineering has no gender. It requires an interest and ability in the physical sciences and mathematics as well as in problem solving. Both men and women have these abilities and interests. People who succeed in engineering have a strong commitment to applying engineering principles to the needs of societies. While women are still underrepresented in engineering, there are many opportunities for them to contribute to solving some of the pressing problems of our times.

STATISTICS ON WOMEN IN ENGINEERING

Currently, women represent only 13 to 15 percent of all engineering graduates. From the late 1960s until the early 1980s, the number of women receiving degrees in science and engineering steadily increased. However, in the mid to late 1980s the number of women receiving degrees in these areas began to decrease.

Studies have shown that the high school years are critical in preparing for a career in engineering. This is particularly true for women and for other underrepresented groups in engineering. It is important for high school women to know what the requirements are for entering college engineering programs. This information is vital when making decisions about what courses to take and which activities to pursue in high school.

The preceding chapters demonstrate that math and physical sciences are basic requirements for any engineering major. Yet a recent study at the University of Michigan found that women in high school continue to take fewer mathematics and science courses than do men. This phenomenon is not due to lack of ability on the part of high school women but, in many cases, it is due to lack of adequate information about the need for these courses in many college majors—particularly in engineering.

Figure 7.1 Women receiving bachelor's degrees in engineering (thousands)

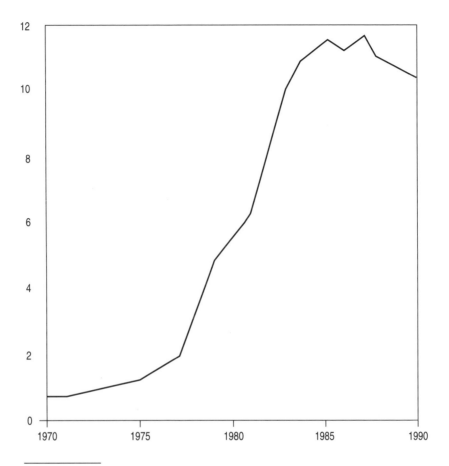

Source: American Association of Engineering Societies

However, high school preparation in math and science does not fully account for the different rates of participation of women and men in engineering. Factors such as self-confidence; stereotypes about women in engineering; encouragement from family, teachers, counselors, and friends; and opportunities for hands-on experiences related to science and engineering play key roles in whether or not a young woman decides to pursue a career in engineering.

ADVICE TO WOMEN ENTERING ENGINEERING

A bachelor's degree in engineering will provide the basic preparation for many exciting opportunities in engineering. However, there are skills that go beyond the basics that are needed to be successful in the field of engineering. Sometimes women overlook these skills and this can be detrimental to their career advancement. These skills fall into four categories: writing, public speaking, teamwork, and getting the job done.

The most important of these career skills is writing ability. Generally, engineers are not known for their writing ability. However, you will need to be able to communicate your work and proposals. Reports, memos, letters, and proposals will be read and evaluated by audiences such as management, clients, peers, and so on.

Learning to tighten your writing so that your ideas are clear and concise will be extremely important. Annual reports, expert testimony, responses to dissatisfied customers, and replies to regulatory agencies are some of the types of writing that will demand versatile writing skills.

The second skill you will need to develop is public speaking. It is often said that public speaking is not an optional skill for engineers. Presentations are a standard part of the engineering work place. Sometimes these speeches are impromptu. Speaking is important because various groups will need to be sold on business and technical merit of your projects.

Many engineers join Toastmasters as a method of improving their public speaking abilities. In fact, at Northwestern University a Toastmasters Club was formed for students and faculty in the school of engineering.

For women engineers, poise and self-confidence when speaking is very important. Avoiding opportunities to speak will not help you improve in this area. Therefore, it is important to join groups and/or take public speaking courses to improve your speaking ability.

The third area in which women need to develop new skills is that of teamwork. Too often engineers can take an "engineering only" outlook. It is easy for engineers to see only the technical aspects of their work. However, it is important for women engineers to look at the broader issues that are facing their companies or organizations.

American industry is increasingly taking a team approach to solving problems. The teams usually are made up of people from diverse disciplines within the organization. The team may consist of a marketing person, an accountant, legal counsel, and an engineer. These cross-functional teams work together to solve a problem or develop a new product. Women engineers need to develop the skills to be active and productive members of these teams.

Finally, women engineers who are entering the work force for the first time need to understand that the work environment and the classroom environment differ greatly. Work assignments can overlap and instructions are not always clear and definitive.

Knowing what it takes to successfully get your job done is something that is not taught in the classroom. Incorporating hands-on experience, such as cooperative engineering education (co-op), in your college career can ease the transition from school to work. As co-op students alternate periods of paid work experience in industry with periods of academic study, they do so in an environment that provides support and encouragement. They learn how to perform as professional engineers in industry before they graduate from college. This type of learning is invaluable to the new engineer.

In addition to developing skills beyond their technical expertise, women in engineering need to be prepared to learn from one another. This can be accomplished through networking and mentoring programs. Whether these programs are formal or informal, they help women "learn the ropes" and further their personal and career development. Knowing your industry and being active in your field of engineering will provide many opportunities for networking and mentoring.

OPPORTUNITIES FOR WOMEN IN ENGINEERING

According to many sources, engineering is one of the best career fields for women. Because women are underrepresented in the engineering work force, many companies aggressively recruit qualified women engineers.

Unlike many other career fields, the salaries of women engineers do not significantly lag behind those of men. In a recent comparison of salaries by the Society of Women Engineers, the average female engineer was making $41,070 while the average male engineer was making $42,888 annually. In order to attract the women engineers, many companies have begun to offer child care, parental leave, and flexible scheduling to accommodate the many roles that women engineers pursue.

SCHOLARSHIP INFORMATION

In addition to the five principal student financial aid programs administered by the federal Department of Education—Pell Grant Programs, Supplemental Educational Opportunity Grants, College-Work Study, National Direct Student Loans, and Stafford Loans (formerly Guaranteed Student Loans)—there are numerous other sources of financial aid for women pursuing undergraduate engineering degrees. Some of these include:

Society of Women Engineers
United Engineering Center, Room 305
345 E. 47th Street
New York, NY 10017

The BPW Foundation Loan for Women in
    Engineering Studies
Business and Professional Women's
    Foundation
2012 Massachusetts Avenue, NW
Washington, DC 20036

Jeanette Rankin Foundation
P.O. Box 4045
Athens, GA 30602

Lutheran Church Women Scholarship
Lutheran Church Women
2900 Queen Lane
Philadelphia, PA 19129

General Federation of Women's Clubs
1734 N. Street, NW
Washington, DC 20036

ORGANIZATIONS FOR WOMEN IN ENGINEERING

Professional Women in Construction
26 Easton Avenue
White Plains, NY 10605

Society of Women Engineers
345 E. 47th Street
New York, NY 10017

ADDITIONAL READING

Beatriz Clewell, *Women of Color in Mathematics, Science, and Engineering: A Review of the Literature* (Washington: Center for Women Policy Studies, 1991).

McIlwee and Robinson, *Women in Engineering: Gender, Power, and The Women in Engineering Directory of College/University Programs,* November, 1991.

Susan Searing, ed., *The History of Women and Science, Health, and Technology: A Bibliographic Guide to the Professions and Disciplines* (Madison, Wisconsin: University of Wisconsin System Women's Studies Library, 1987).

Betty Vetter, *Professional Women and Minorities: A Total Human Resource Data Compendium* (Washington, DC, Commission on Professionals in Science and Technology; June, 1992).

Workplace Culture, SUNY Press, Albany, NY, 1992.

The Woman Engineer (a magazine for women in engineering), Equal Opportunity Publications, 44 Broadway, Greenlawn, NY 11740.

U.S. Woman Engineer (the magazine of the Society of Women Engineers), The Society of Women Engineers, 345 East 47th Street, New York, NY 10017.

The Catalogue of Resource Materials, an extensive database of Women in Engineering Programs funded by the National Science Foundation.

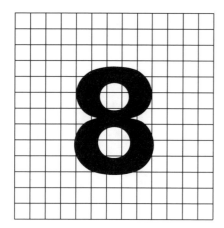

MINORITIES IN ENGINEERING

Did you know that the innovative body design of the Saturn SL I automobile was the responsibility of an African-American engineer? Did you know that the manager of research at one of IBM's major research laboratories is a Hispanic engineer? Did you know that a major research laboratory, which receives money from the University of California and U.S. Department of Energy, has an American Indian Program and a joint engineering pact with the Navajos?

STATISTICS ON MINORITIES IN ENGINEERING

While these achievements are significant, they do not mean that minorities have achieved equality in the field of engineering. Minorities represent approximately 22 percent of our total society. There are over 54 million African-Americans, Hispanics, American Indians, and Alaskan natives. However, these groups are underrepresented in engineering. According to the National Action Council for Minorities in Engineering the trend for minorities in engineering is up. However, representation for all minority groups except Asian-Americans is still well below their proportions in the general public.

LeRoy Callender, P.E., and President of LeRoy Callender PC Consulting Engineers based in New York, offers a partial explanation for the underrepresentation of African-Americans. "A few years ago, there weren't many Blacks majoring in engineering because it didn't make sense to work hard for a degree and not be able to get a job. Now, a Black engineer is worth his weight in gold . . ."

The Council on Competitiveness would agree with Mr. Callendar. The council estimates that by the year 2000 the United States may have a shortage of 500,000 engineers and scientists. This anticipated shortage is

Figure 8.1 Minority Engineering Freshmen 1972–1990

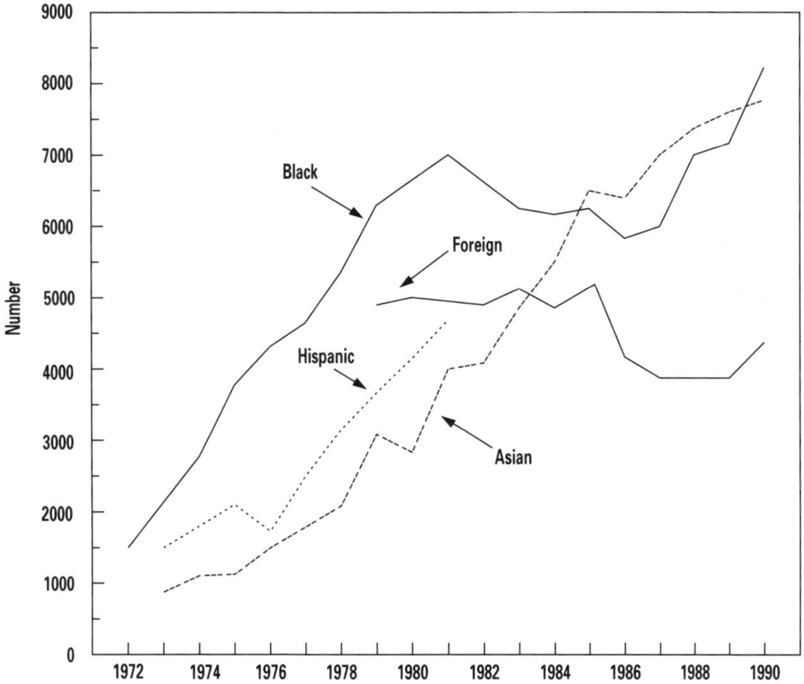

Data Source: Engineering Manpower Commission

Graph by CPST for *Manpower Commission*, Vol 28, No. 3, April/May 1991, p. 20.

the result of too few people in the engineering education "pipeline" and is causing companies and schools to target minority engineering students for scholarships and permanent job offers after graduation. In other words, Mr. Callendar's assessment of the current opportunities for minority engineers is correct!

A more serious barrier to minorities entering engineering careers is inadequate educational preparation. When students do not take a sufficient amount of math and physical science during high school, they have an educational deficit that is extremely difficult, if not impossible, to overcome at the college level.

ADVICE FOR MINORITY ENGINEERS

If you plan to pursue a career in engineering, it is important to be prepared. During high school you need to take as much mathematics and science as possible. This means that you should take calculus and trigonometry in addition to the usual algebra and geometry. You also need to

Figure 8.2 Engineering Enrollments, Fall 1990

ENGINEERING ENROLLMENTS, FALL 1990							
UNDERGRADUATE STUDENTS			GRADUATE STUDENTS				
		All Years		MS/PE	PhD	All Grad Students	
	1st Year	Full time	Part time	Degree	Degree	Full time	Part time
All Students	94,346	338,842	41,445	42,415	30,041	72,456	45,378
Women	16,674	55,915	5,901	6,236	3,522	9,758	6,853
Black	8,370	20,909	2,653	760	326	1,086	948
Hispanic	5,885	19,483	2,118	940	336	1,276	925
Nat. American	526	1,468	139	87	26	113	111
Asian	7,926	31,257	3,953	2,773	1,980	4,753	4,257
Foreign Nat.	4,442	20,562	2,006	16,787	15,469	32,256	6,891
Percent							
All Students	100.00	100.00	100.00	100.00	100.00	100.00	100.00
Women	17.67	16.50	14.24	14.70	11.72	13.47	15.10
Black	8.87	6.17	6.40	1.79	1.09	1.50	2.09
Hispanic	6.24	5.75	5.11	2.22	1.12	1.76	2.04
Nat. American	0.56	0.43	0.34	0.21	0.09	0.16	0.24
Asian	8.40	9.22	9.54	6.54	6.59	6.56	9.38
Foreign Nat.	4.71	6.07	4.84	39.58	51.49	44.52	15.19

Data Source: Engineering Manpower Commission

Manpower Commission, Vol 28, No. 3, April/May 1991, p. 20.

take courses in biology, chemistry, and physics, including advanced courses in chemistry and physics. These courses will prepare you for the basic educational requirements for most engineering majors.

With the right educational tools and a strong desire to achieve, anyone can make an impact. However, minorities in engineering have had few role models or mentors. Therefore, finding support for your interest in engineering is very important.

Your family, your teachers, and your guidance counselors are people who can possibly help you identify individuals, organizations, and local industries knowledgeable about engineering *and* willing to be supportive of your interest in engineering. If these groups are unable to put you in touch with other individuals who can assist you, contact the organizations listed at the end of this chapter.

Figure 8.3 Gains in College Enrollments, 1986 to 1988

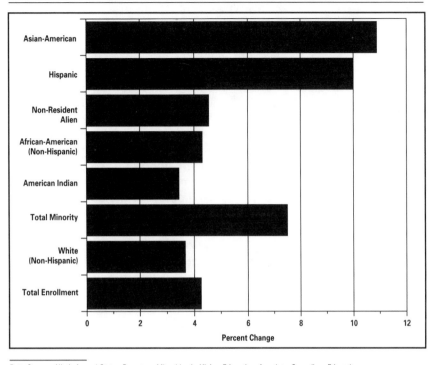

Data Source: *Ninth Annual Status Report on Minorities in Higher Education*, American Council on Education
Manpower Commission, Vol 28, No. 2, April/May 1991, p. 18.

Organizations such as the National Society of Black Engineers (NSBE) and the Society for Hispanic Professional Engineers (SHPE) seek to identify minority students and encourage them to pursue education and employment in engineering. These organizations offer such things as tutoring programs, group study sessions, juniors and senior high school outreach programs, technical seminars and workshops, a national communications networks, resume books, career fairs, and so on.

Local industry, business, and government agencies that employ engineers would also be happy to assist you in meeting engineers and showing you what engineers do in their setting. Often a phone call to the human resources department can be very helpful in setting up such a meeting.

Selecting an engineering school will be an important step in your engineering career. When evaluating colleges and universities, it is not only important to examine the curriculum and learn about the faculty, it is also important to meet members of the local chapters of NSBE, SHPE, and other minority engineering organizations on campus. Many chapters are extremely active and provide a strong support network for their members. Make sure the schools you are looking at have strong chapters of this type of organization.

During your college career it is also important to gain hands-on experience related to engineering. This may be through summer work experience. But in college it is important to take a serious look at cooperative engineering education (co-op) programs.

Co-op is a well known educational approach for engineers, and millions of engineers have graduated from colleges with co-op experience. Co-op is the type of hands-on experience that industry values. In 1991, major corporations that recruit engineers began saying that they will make their permanent offers to their co-op students first. Therefore, it will be increasingly important for minority engineering students to participate in co-op programs so that they are assured of being part of the pool of job candidates for permanent employment.

You and your family should talk to the staff of the co-op office at any college or university that you are considering attending. Sometimes there is a misconception that college co-op programs are not for the serious academic student. This is far from the truth. Co-op employers are seeking the best and the brightest and they have a vested interest in their co-op students receiving their degrees. A reputable co-op program will not continue to work with employers who hire students away from the classroom. Therefore, it is important for you and your family to visit this office and ask lots of questions.

SALARY STATISTICS FOR MINORITY ENGINEERS

Engineering and computer science offer some of the highest starting salaries and the College Placement Council indicates that salary offers for most types of engineers is expected to rise. In addition, pay for engineers and scientists remains relatively high throughout their careers. The largest salaries generally go to those in management and supervisory positions in engineering.

ORGANIZATIONS FOR MINORITY ENGINEERS

American Indian Science and Engineering
  Society
1630 30th Street, Suite 301
Boulder, CO 80301-1014
(303) 492-8658

National Action Council for Minorities in
  Engineering
3 West 35th Street
New York, NY 10001
(212) 279-2626

Figure 8.4 Average Salary Offers

AVERAGE STARTING SALARY OFFERS TO BACHELOR'S GRADUATES, SELECTED FIELDS, MARCH 1991			
Engineering		Accounting	$26,603
Aerospace	$30,318	Business Admin	23,096
Chemical	36,990	Bus. Econ., Finance	25,002
Civil	29,798	Communications	21,583
Electrical	32,943	Elem. Education	19,937
Mechanical	33,803	Psychology	18,802
Computer Science	30,153	Nursing	27,964
Info Sciences/Systems	28,800	Allied Health	27,774
Chemistry	27,625	Biological Sciences	20,410
Mathematics	25,595	Physics	28,176

Source: College Placement Council

Data Source: College Placement Council

Manpower Commission, Vol 28, No. 3, April/May 1991, p. 15.

National Association of Minority
  Engineering
College of Engineering
202 West Boyd, CEC 107
Norman, OK 73109
(405) 325-4161

National Society of Black Engineers
1454 Duke Street
Alexandria, VA 22314
(703) 549-2207

Young Black Programmers Coalition
P.O. Box 11243
Jackson, MO 39213
(601) 634-5775

OTHER SOURCES OF INFORMATION

The National Science Foundation has established a set of 12 comprehensive regional centers for minorities. These centers carry out a wide range of programs to improve teaching, interest youngsters, and bring parents and counselors into the picture.

The Alliances for Minority Participation Program is another NSF program designed to increase the participation of minorities in engineering and science careers. The goal of this program is to increase the number of bachelor's degrees in the natural sciences and engineering from 14,000 to 50,000 by the year 2000. This will be accomplished through coalitions of universities, industries, states, and cities.

Directory of Career Resources for Minorities, Ready Reference Press, P.O. Box 5169, Santa Monica, CA 90405.

Directory of Special Programs for Minority Group Members: Career Information Services, Employment Skills Banks, Financial Aid Sources, Garrett Park Press, P.O. Box 190F, Garrett Park, MD 20896.

ADDITIONAL READING

Beatriz Clewell, *Women of Color in Mathematics, Science, and Engineering: A Review of the Literature* (Washington: Center for Women Policy Studies, 1991).

Sheila Humphreys, *Women and Minorities in Science,* Westview Press, 1982.

Jane Butler Kahle, *Double Dilemma: Minorities and Women in Science Education,* (West Lafayette, IN: Purdue University Press, 1982).

Rebecca Rawls, "Minorities in Science", *Chemical and Engineering News,* American Chemical Society, Washington, DC, April 15, 1991.

Magazines:

NSBE Magazine: National Society of Black Engineers, 1454 Duke Street, Alexandria, VA 22314.

NSBE Bridge National Society of Black Engineers, 1454 Duke Street, Alexandria, VA 22314.

US Black Engineer, Career Communications Group, Inc., 729 East Pratt Street, Suite 504, Baltimore, MD 21202.

Hispanic Engineer Magazine, Career Communications Group, Inc., 729 East Pratt Street, Suite 504, Baltimore, MD 21202.

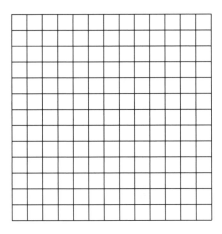

APPENDIX A
ENGINEERING SOCIETIES

American Society of Body Engineers
Wilshire Office Center, Suite 3031
24634 Five Mile Road
Redford, MI 48239

Society of Automotive Engineers (SAE)
400 Commonwealth Drive
Warrendale, PA 15096

American Ceramic Society
757 Brooks Edge Plaza
Westerville, OH 43081

American Institute of Chemical Engineers
345 East 47th Street
New York, NY 10017

American Society of Naval Engineers
1452 Duke Street
Alexandria, VA 22314

American Society of Mechanical Engineers
345 East 47th Street
New York, NY 10017

The American Chemical Society
1155 Little Falls Street
Falls Church, VA 22041

ASM International (metallurgy)
American Society for Metals
Metals Park, OH 44073

Association of Iron and Steel Engineers
Three Gateway Center, Suite 2350
Pittsburgh, PA 15222

Society of Petroleum Engineers
P.O. Box 833836
Richardson, TX 75083

Society of Professional Well Log Analysts
6001 Gulf Freeway, Suite C129
Houston, TX 77023

Society of Plastics Engineers
14 Fairfield Drive
Brookfield Center, CT 06085

National Association of Power Engineers
2350 East Devon, Suite 115
Des Plaines, IL 60018

American Society of Safety Engineers
1800 East Oakton Street
Des Plaines, IL 60016

Society of Die Casting Engineers
2000 Fifth Avenue
River Grove, IL 60171

American Institute of Mining, Metallurgical
and Petroleum Engineers
345 East 47th Street
New York, NY 10017

Society of Mining Engineers
Caller No. D
Littleton, CO 80127

American Nuclear Society
555 North Kensington Avenue
La Grange Park, IL 60525

The Junior Engineering Technical Society
(JETS)
National Office
345 East 47th Street
New York, NY 10017

American Institute of Plant Engineers
3975 Erie Avenue
Cincinnati, OH 45208

American Society of Civil Engineers
(ASCE)
345 East 47th Street
New York, NY 10017

American Society of Heating, Refrigeration
and Air Conditioning Engineers (ASHRAE)
1791 Tullie Circle, NE
Atlanta, GA 30329

American Society of Lubrication Engineers
838 Busse Highway
Park Ridge, IL 60068

American Society of Plumbing Engineers
(ASPE)
3617 Thousand Oaks Boulevard, #210
Westlake, CA 91362

Institute for the Advancement of
Engineering (IAE)
P.O. Box 1305
Woodland Hills, CA 91365

Institute of Industrial Engineers
25 Technology Park
Norcross, GA 30092

Insulated Cable Engineers Association
(ICEA)
P.O. Box P
Yarmouth, MA 02664

International Association of Cylindrical
Hydraulic Engineers
37 Ridgeview Drive
Chicksaw, AL 36611

National Association of Corrosion
Engineers
P.O. Box 218340
Houston, TX 77218

National Society of Architectural Engineers
P.O. Box 395
Lawrence, KS 66044

National Institute of Ceramic Engineers
757 Brookside Plaza Dr.
Westerville, OH 43081-2821

Society of Motion Picture and Television
Engineers
595 West Hartsdale Avenue
White Plains, NY 10607

Society of Fire Protection Engineers
60 Batterymarch Street
Boston, MA 02110

American Society of Bakery Engineers
Two Riverside Plaza, Room 1921
Chicago, IL 60606

Society of Logistics Engineers (SOLE)
125 West Park Loop Street, 210
Huntsville, AL 35806

Illuminating Engineering Society of North
America
345 East 47th Street
New York, NY 10017

Society of Manufacturing Engineers
One SME Drive
P.O. Box 930
Dearborn, MI 48128

American Society of Agricultural Engineers
2950 Niles Road
St. Joseph, MI 49085

National Engineering Consortium
505 North Lake Shore Drive, Suite 4808
Chicago, IL 60611

Association of Energy Engineers
4025 Pleasantdale Road, Suite 420
Atlanta, GA 30340

American Society for Hospital Engineering
(ASHE)
American Hospital Association
840 North Lake Shore Drive
Chicago, IL 60611

American Academy of Environmental
Engineers (AAEE)
132 Holiday Court, #206
Annapolis, MD 21401

Institute of Electrical and Electronics
Engineers (IEEE)
345 East 47th Street
New York, NY 10017

International Society for Optical
Engineering (SPIE)
P.O. Box 10
1022 Nineteenth Street
Bellingham, WA 98227

Society of Packaging and Handling
Engineers
Reston International Center
Reston, VA 22091

Pattern Recognition Society
c/o National Biomedical Research
Foundation
Georgetown Medical Center
3900 Riverside Road, NW
Washington, DC 20007

American Society of Sanitary Engineering
P.O. Box 40362
Bay Village, OH 44140

National Association of Radio and
Telecommunications Engineers
P.O. Box 15029
Salem, OR 97309

Institute of Transportation Engineers
525 School Street, SW, Suite 410
Washington, DC 20024

American Institute of Aeronautics and
Astronautics
370 L'Enfant Promenade, SW
Washington, DC 20024-2518

Society of Flight Test Engineers
P.O. Box 4047
Lancaster, CA 93539

American Society of Cost Engineers
308 Monongahela Building
Morgantown, WV 26505

INTERDISCIPLINARY PROFESSIONAL AND TRADE ASSOCIATIONS

American Association for the Advancement
of Science
1515 Massachusetts Avenue, NW
Washington, DC 20005

American Association of Engineering
Societies Inc.
345 East 47th Street
New York, NY 10007

Association for Women in Science
1346 Connecticut Avenue, NW
Washington, DC 20036

National Consortium for Graduate Degrees
for Minorities in Engineering Inc.
P.O. Box 537
Notre Dame, IN 46556

National Society of Professional Engineers
1420 King Street
Alexandria, VA 22314

Scientific Manpower Commission
1500 Massachusetts Avenue, NW, Suite 381
Washington, DC 20005

COMPUTER SCIENCE

American Federation of Information
Processing Societies Inc.
1815 North Lynn Street
Arlington, VA 22209

American Society for Information Science
1010 Sixteenth Street, NW
Washington, DC 20036

Association for Computational Linguistics
SRI International
Menlo Park, CA 94025

Association for Computing Machinery Inc.
1133 Avenue of the Americas
New York, NY 10036

Association for Educational Data Systems
1201 Sixteenth Street, NW
Washington, DC 20036

Association for Systems Management
24587 Bagley Road
Cleveland, OH 44138

Association of Computer Programmers and
Analysts
11800 Sunrise Valley Drive
H808
Reston, VA 22091

MATHEMATICS

American Mathematical Society
P.O. Box 6248
Providence, RI 02940

American Statistical Association
806 Fifteenth Street, NW, Suite 640
Washington, DC 20005

Mathematical Association of America
1529 Eighteenth Street, NW
Washington, DC 20036

Operations Research Society of America
428 East Preston Street
Baltimore, MD 21202

Association of Data Processing Service
Organizations Inc.
1300 North Seventeenth Street
Arlington, VA 22209

Data Processing Management Association
505 Busse Highway
Park Ridge, IL 60068

Independent Computer Consultants
Association
Box 27412
St. Louis, MO 63141

Operations Research Society of America
428 East Preston Street
Baltimore, MD 21202

Society for Computer Simulation
P.O. Box 2228
La Jolla, CA 92038

Society of Actuaries
208 South LaSalle Street
Chicago, IL 60604

Society for Industrial and Applied
Mathematics
1405 Architects Building
117 South Seventeenth Street
Philadelphia, PA 19103

PHYSICAL SCIENCES

American Association of Petroleum
Geologists
P.O. Box 979
Tulsa, OK 74101

American Association of Physicists in
Medicine
335 East 45th Street
New York, NY 10017

American Association of Textile Chemists
and Colorists
P.O. Box 12215
Research Triangle Park, NC 27709

American Astronomical Society
1816 Jefferson Place, NW
Washington, DC 20036

American Congress on Surveying and
Mapping
5410 Grosvenor Lane
Bethesda, MD 20814

American Geological Institute
5205 Leesburg Pike
Falls Church, VA 22041

American Geophysical Union
2000 Florida Avenue, NW
Washington, DC 20009

American Institute of Chemists Inc.
7315 Wisconsin Avenue
Bethesda, MD 20014

American Institute of Physics Inc.
335 East 45th Street
New York, NY 10017

American Meteorological Society
45 Beacon Street
Boston, MA 02108

American Physical Society
335 East 45th Street
New York, NY 10017

American Society of Biological Chemists
Inc.
9650 Rockville Pike
Bethesda, MD 20014

Association of Consulting Chemists and
Chemical Engineers Inc.
50 East 41st Street
New York, NY 10017

Biophysical Society of America
9650 Rockville Pike, Room 404
Bethesda, MD 20014

WOMEN AND MINORITIES

American Indian Science and Engineering
Society
1630 30th Street
Suite 301
Boulder, CO 80301-1014

Association for Women in Computing
P.O. Box 2293
Grand Central Station
New York, NY 10163

OTHER

Accreditation Board for Engineering and
Technology
345 East 47th Street
New York, NY 10017-2397

National Council of Examiners for
Engineering
P.O. Box 1686
Clemson, SC 29633-1686

National Commission for Cooperative
Education
P.O. Box CAR
360 Huntington Avenue
Boston, MA 02115

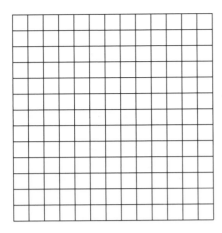

APPENDIX B
A GUIDE TO FINANCIAL AID

Cost alone should not deter you from obtaining an engineering education. While you and your family are expected to meet some of the costs of higher education, there are financial assistance programs to help you make the dream of an engineering degree a reality.

The college or university of your choice will have a representative on staff whose sole responsibility is to assist students in making application for all of the available financial aid programs they qualify for. These representatives can assist you in filling out the various forms and will answer your questions concerning the types and amounts of assistance available.

It is very important to understand the differences among the five types of aid available. This aid can take the form of:

- scholarships
- grants
- loans
- work-study job
- co-op

The last two types of aid are relatively recent innovations. A discussion of them concludes this appendix.

SCHOLARSHIPS

A scholarship is a monetary grant to a student who qualifies under a variety of circumstances. There are scholarships for the academically gifted. Other scholarships are based on special abilities such as musical talent, athletics, speech, and drama. Occasionally, the scholarship has two conditions that must be met: ability *and* financial need.

The key point to remember is that a scholarship is a grant, or gift, of money for a specific purpose (education) with specific requirements (talent, academic merit, and/or financial need). It does not require repayment. Each institution has its own requirements and qualifications for scholarships.

GRANTS

A grant is a gift of financial aid for a specific purpose (education). Like a scholarship, it may impose conditions on the recipient and does not require repayment. Sources of grants include the institution itself; local, state, and federal government; and private industry and groups. Each grant has specific conditions and requirements that must be met.

LOANS

A loan is not a grant or a scholarship. It is a legal obligation that must be repaid. A student loan can be obtained from the institution, from a bank or savings and loan, or from a government agency. The government agency's role is usually limited to insuring or guaranteeing the lender—the school, a bank, or a savings and loan—that you will repay the loan according to the terms of the contract.

Certain loans, especially those for students in high-demand, low-supply career fields, often have special clauses permitting portions of the loan to be cancelled if conditions are met. An example is a loan made to an education major. According to one such agreement, a fixed percentage of the loan's total is cancelled for each year the graduate teaches in an inner-city school. There are similar arrangements for medical students as well as for those entering the U.S. Public Health Service.

The important thing to keep in mind is that this assistance is a loan, which must be repaid. Interest charges are added to the loan. Repayment usually does not begin until after you have graduated, and you are obligated to repay the loan in a specific time period. If you default on the loan, the lender may take legal action against you to ensure that the loan is repaid. If the federal government guarantees your loan, and you fail to pay, then legally your income tax refund checks and your wages can be garnisheed until the debt is satisfied.

APPLYING FOR FINANCIAL AID

Schools use a confidential statement of income and assets from the parents and student to determine what financial aid a student is eligible for. If the student is independent and does not live at home with his or her parents, then the application will reflect information about the student alone. If you fall under this category, you might want to discuss the options available to you with the financial aid representative before proceeding any further.

Application forms, the confidential statement of assets and income, can vary from one school to another. You should determine exactly which form the school of your choice requires and be certain to submit only that form. The forms have a deadline for submission, so you should be sure that you have the form and collect the information in plenty of time to meet this deadline.

Questions on the application will assess the financial condition of the family unit. From your responses, the financial aid committee can determine an estimate of how much money your family can reasonably be expected to contribute toward your education and how much outside assistance you will require.

This assessment will take into account your family's size, income, assets and liabilities, and the number of family members attending postsecondary educational facilities. Your parents' age and the resources available to you will also enter into the final analysis.

A standard formula is used, to ensure that all students are considered in an impartial manner. Your financial aid administrator should be made aware of any special family circumstances that should be considered in this formula along with the information solicited on the application. This special information should be detailed in a letter, which should be sent with any supporting documentation to the aid representative.

HOW MUCH WILL MY EDUCATION COST?

What a four-year degree costs depends on the type of institution you select. Private, Ivy League colleges cost a great deal more than local, state-supported colleges. Each school will, on request, provide you with a breakdown of its student expenses. This budget consists of the following items:

- Tuition and fees—the amount you will be charged for your classes.
- Books and supplies—This will vary based on the type of academic program you have selected. Be sure to ask about special books or equipment (lab supplies and computers) required of students in your program.
- Room and/or housing—This amount will reflect either rates for available quarters on campus (dorms) or, if no quarters are available, the prevailing rents in the local community.
- Board and meals—the cost you will have to pay for your daily meals. If there is a school cafeteria, the amount will be somewhat fixed. If you must eat out, the cost can vary substantially. Sometimes a fixed board fee will include laundry and telephone. Be sure to ask.
- Personal expenses—This is your allowance, a fixed amount of money for entertainment, personal items, and miscellaneous charges you might incur.

- Transportation—This is an allowance for trips from school back to your home, for example.

While these costs can vary, especially if you elect to live at home and attend a local college, these factors must be considered to determine what your college education will really cost. You might want to fill out the following chart to help you determine the basic costs.

Estimating the Cost of Your Education

Tuition and Fees	$_____
Books and Supplies	_____
Room/Housing	_____
Board/Meals	_____
Personal Expenses	_____
Transportation	_____
Other Expenses	_____
Estimated College Cost	_____
Estimated Family Contribution	_____
Estimated Aid Requirement	_____

HOW WILL I KNOW WHAT AID I'LL RECEIVE?

The financial aid office at your school will notify you of its decision in what is called an award letter or award notification form. This will tell you what kind or kinds of aid you can expect and the amounts you are eligible for.

This will cover all of the available forms of aid you have qualified for, including scholarships, grants, student employment, and loans. Many institutions call this award a financial aid package.

After you have received this letter, you should notify the financial aid representative that you either accept or decline the aid. Some schools will withdraw the offer of assistance if you do not notify them by a specified deadline.

WORK-STUDY JOB AND CO-OP

Although these two forms of assistance do not fall under the traditional categories, they may be available to you even if you do not qualify for any other assistance. The work-study job assigns you to a specific job and pays you for every hour you work. This can be used to supplement scholarships and grants as long as you ensure that your work schedule does not conflict with your education.

A co-op adds at least a year to your college life, but it can be of substantial benefit and actually put you ahead of students who were able to attend school full time. A co-op is a program of school and work. You are assigned to a company where you will work in an area closely related to your chosen profession. You work certain hours and attend school certain hours. In doing so, you will gain valuable experience in your field that can make a financial difference when you graduate and pursue your first job.

Remember, there is no reason not to get an education. Financial aid programs can help you afford the cost of higher education.

FINANCIAL AID BIBLIOGRAPHY

Applying for Financial Aid: Financial Aid Services
American College Testing Program
P.O. Box 168
Iowa City, IA 52243

Chronicle Student Aid Annual
Chronicle Guidance Publications
Aurora Street
Moravia, NY 13118

Directory of Financial Aids for Minorities
Reference Service Press
3540 Wilshire Boulevard, Suite 310
Los Angeles, CA 90010

Financial Aid for Higher Education and Undergraduate Catalog
William C. Brown Publishers
2460 Kerper Boulevard
Dubuque, IA 52001

Guide to Academic Scholarships
Octameron Press
P.O. Box 3437
Alexandria, VA 22302

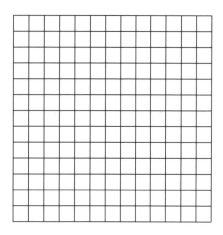

APPENDIX C
ENGINEERING FRATERNITIES

Alpha Pi Mu (industrial engineering)
P.O. Box 934
Blacksburg, VA 24060

Chi Epsilon (civil engineering)
169 Engineering Research Center
Texas A&M
College Station, TX 77843

Eta Kappa Nu (electrical engineering)
Department of Electrical Engineering
155 Electrical Engineering Building
University of Illinois
Urbana, IL 61801

Kappa Eta Kappa (electrical engineering)
114 North Orchard Street
Madison, WI 53715

Keramos (ceramic engineering)
c/o Materials Science and Engineering
Department
110 Engineering Annex
Iowa State University
Ames, IA 50011

Omega Chi Epsilon (chemical engineering)
Department of Chemical Engineering
McNeese State University
Lake Charles, LA 70609

Phi Kappa Upsilon (engineering and
architecture)
21000 West Nine Mile Road
Southfield, MI 48075

Phi Tau Sigma (mechanical engineering)
Department of Mechanical Engineering
Tennessee Technological University
Cookeville, TN 38505

Sigma Gamma Tau (aerospace engineering)
Department of Aerospace Engineering and
Engineering Mechanics
University of Texas at Austin
Austin, TX 78712

Sigma Phi Delta (engineering)
438 Smithfield Street
East Liverpool, OH 43920

Tau Alpha Pi (engineering technology)
P.O. Box 266
Riverdale, NY 10471

Tau Beta Pi Association (engineering)
P.O. Box 8840
University Station
Knoxville, TN 37996

Theta Tau (engineering)
9974 Old Olive Street Road
St. Louis, MO 63141

Kappa Theta Epsilon (cooperative education
honor society)
c/o Cooperative Education Program
Virginia Tech
252 Henderson Hall
Blacksburg, VA 24061

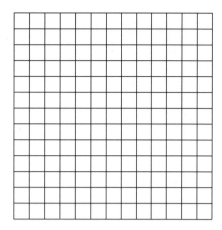

APPENDIX D

RECOMMENDED READING

Where to Get More Information

There are many sources of information on engineering careers. Free and low-cost literature on specific engineering careers is available from professional engineering societies as well as government agencies, corporations, and educational institutions.

Career information on engineering is available in libraries, career centers, and guidance offices. State occupational information coordinating committees can provide information about engineering opportunities in specific states and localities.

Personal contacts are also an excellent source of information on careers in engineering. Talking to professional engineers about what they do can be very helpful in career decision making. Local chapters of professional engineering societies and local industries can often provide the names of engineers and engineering technologists who are available to speak about their careers.

Beakley, George C., *Careers in Engineering and Technology* (New York: Macmillan, 1987).

Beakley, George C., *Engineering: An Introduction to a Creative Profession* (New York: Macmillan, 1986).

Grundfest, Sandra, ed., *Peterson's Annual Guide to Job Opportunities for Engineers, Computer Scientists and Physical Scientists* (Princeton: Peterson's Guides, 1992).

Vetter, Betty M., *Opportunities in Science and Engineering* (Washington, DC: Commission on Professionals in Science and Technology, 1984).

Woodburn, John H., *Opportunities in Energy Careers* (Lincolnwood, IL: VGM Career Horizons, 1992).

Wright, John W. and Edgard J. Dwyer, *The American Almanac of Jobs and Salaries 1990–1991 Edition* (New York: Avon Books, 1990).

VGM CAREER BOOKS

CAREER DIRECTORIES
Careers Encyclopedia
Dictionary of Occupational
Titles
Occupational Outlook
Handbook

CAREERS FOR
Animal Lovers
Bookworms
Computer Buffs
Crafty People
Culture Lovers
Environmental Types
Film Buffs
Foreign Language Aficionados
Good Samaritans
Gourmets
History Buffs
Kids at Heart
Nature Lovers
Night Owls
Number Crunchers
Plant Lovers
Shutterbugs
Sports Nuts
Travel Buffs

CAREERS IN
Accounting; Advertising;
Business; Child Care;
Communications; Computers;
Education; Engineering;
the Environment; Finance;
Government; Health Care;
High Tech; Journalism; Law;
Marketing; Medicine;
Science; Social &
Rehabilitation Services

CAREER PLANNING
Admissions Guide to Selective
Business Schools
Beating Job Burnout
Beginning Entrepreneur
Career Planning &
Development for College
Students & Recent Graduates
Career Change

Careers Checklists
Cover Letters They Don't
Forget
Executive Job Search Strategies
Guide to Basic Cover Letter
Writing
Guide to Basic Resume Writing
Guide to Temporary
Employment
Job Interviews Made Easy
Joyce Lain Kennedy's Career
Book
Out of Uniform
Resumes Made Easy
Slam Dunk Resumes
Successful Interviewing for
College Seniors
Time for a Change

CAREER PORTRAITS
Animals	Nursing
Cars	Sports
Computers	Teaching
Music	Travel

GREAT JOBS FOR
Communications Majors
English Majors
Foreign Language Majors
History Majors
Psychology Majors

HOW TO
Approach an Advertising
Agency and Walk Away with
the Job You Want
Bounce Back Quickly After
Losing Your Job
Choose the Right Career
Find Your New Career Upon
Retirement
Get & Keep Your First Job
Get Hired Today
Get into the Right Business
School
Get into the Right Law School
Get People to Do Things Your
Way
Have a Winning Job Interview

Hit the Ground Running in
Your New Job
Improve Your Study Skills
Jump Start a Stalled Career
Land a Better Job
Launch Your Career in TV
News
Make the Right Career Moves
Market Your College Degree
Move from College into a
Secure Job
Negotiate the Raise You
Deserve
Prepare a Curriculum Vitae
Prepare for College
Run Your Own Home Business
Succeed in College
Succeed in High School
Write a Winning Resume
Write Successful Cover Letters
Write Term Papers & Reports
Write Your College Application
Essay

OPPORTUNITIES IN
This extensive series provides
detailed information on nearly
150 individual career fields.

RESUMES FOR
Advertising Careers
Banking and Financial Careers
Business Management Careers
College Students &
Recent Graduates
Communications Careers
Education Careers
Engineering Careers
Environmental Careers
50 + Job Hunters
Health and Medical Careers
High School Graduates
High Tech Careers
Law Careers
Midcareer Job Changes
Sales and Marketing Careers
Scientific and Technical Careers
Social Service Careers
The First-Time Job Hunter

VGM Career Horizons
a division of *NTC Publishing Group*
4255 West Touhy Avenue
Lincolnwood, Illinois 60646-1975